Worldviews
and Ecology

ECOLOGY AND JUSTICE

An Orbis Series on Global Ecology

Advisory Board Members
Mary Evelyn Tucker
John A. Grim
Leonardo Boff
Sean McDonagh

The Orbis Series *Ecology and Justice* publishes books that seek to integrate an understanding of Earth as an interconnected life system with concerns for just and sustainable systems that benefit the entire Earth. Books in the Series concentrate on ways to:

- reexamine the human-earth relationship in light of contemporary cosmological thought
- develop visions of common life marked by ecological integrity and social justice
- expand on the work of those who are developing such fields as eco-theology, ecojustice, environmental ethics, ecofeminism, deep ecology, social ecology, bioregionalism, and animal rights
- promote inclusive participative strategies that enhance the struggle of the Earth's voiceless poor for justice
- deepen appreciation for and expand dialogue between religious traditions on the issue of ecology
- encourage spiritual discipline, social engagement, and the reform of religion and society toward these ends.

Viewing the present moment as a time for responsible creativity, the Series seeks authors who speak to ecojustice concerns and who bring into dialogue perspectives from the Christian community, from the worlds' other religions, from secular and scientific circles, and from new paradigms of thought and action.

Also in the Series

John B Cobb, Jr., *Sustainability: Economics, Ecology, and Justice*
Charles Pinches and Jay B. McDaniel, editors, *Good News for Animals?*
Frederick Ferré, *Hellfire and Lightning Rods*
Ruben L. F. Habito, *Healing Breath: Zen Spirituality for a Wounded Earth*
Eleanor Rae, *Women, the Earth, the Divine*

ECOLOGY AND JUSTICE SERIES

Worldviews and Ecology

Religion, Philosophy, and the Environment

Edited by
MARY EVELYN TUCKER
and
JOHN A. GRIM

ORBIS BOOKS
Maryknoll, New York 10545

Fourth Printing, January 1999

The Catholic Foreign Mission Society of America (Maryknoll) recruits and trains people for overseas missionary service. Through Orbis Books, Maryknoll aims to foster the international dialogue that is essential to mission. The books published, however, reflect the opinions of their authors and are not meant to represent the official position of the society.

Library of Congress Cataloging-in-Publication Data

Worldviews and ecology : religion, philosophy, and the environment /
    edited by Mary Evelyn Tucker and John A. Grim.
        p.   cm. — (Ecology and justice series)
    ISBN 0-88344-967-6 (pbk. : alk. paper)
    1. Nature—Religious aspects.   2. Human ecology—Religious
aspects.   3. Religions.   I. Tucker, Mary Evelyn.   II. Grim, John.
III. Series: Ecology and justice.
BL65.N35W67   1994
291.1'78362—dc20   94-21625
                                                                CIP

To Thomas Berry and Brian Swimme

This Orbis Books edition of *Ecology and Worldviews* is dedicated with gratitude to Thomas Berry, who has shown us the spiritual and historical contributions of the various world religions and has brought us to an appreciation of the significance of many of the newly emerging worldviews represented in this volume. It is also dedicated with joy to Brian Swimme, who with his unique sense of awe has opened up our participation in the Universe story by his extraordinary blending of native, scientific, and poetic voices. Let the celebration of the ecozoic age begin!

Acknowledgements

We give special thanks to Dorothy Baumwoll, Pauline Fletcher, and Steven Styers for their assistance in bringing this volume to publication. They have been enormously kind, patient, and efficient, and it has been a pleasure to work with them.

Mary Evelyn Tucker
John A. Grim

Contents

Contemporary Ecological Perspectives

Preface

Mary Evelyn Tucker
John A. Grim

The reality of the ecological crisis assaults us from many directions. The scale and complexity of the problems and the elusiveness of long-range solutions that the media make us conscious of have become increasingly hard to ignore. So we know that we would be negligent if we did not continue to seek for ways of finding paths out of this maze of ongoing environmental degradation. Many perspectives, including those arising from religion and philosophy, will be necessary in the important task of rethinking human-earth relations. Exploring some of these is what this volume is about.

Certainly the world religions have been instrumental in formulating views of nature and in creating perspectives on the role of the human in nature. It should be clear, then, that the examination of different religious worldviews may be critical in the task of analyzing the roots of the environmental crisis as well as in proposing solutions.

In addition to valuable insight from the Western religious traditions, there is much to be learned from Asian religions and from Native traditions. Each of these areas are beginning to be explored by scholars and practitioners who are searching for new models of ecological wholeness and reciprocity. For some traditions, such as the Western religions, the question will be how can appreciation for nature be more fully integrated into a religious context, while for others, such as Native traditions, where the earth *is* the central religious context, the question will be how to articulate that knowledge in relation to modernity. All these efforts are key to formulating more inclusive worldviews and ethics necessary for solving the environmental crisis.

The premise of this book, however, is that *no one religious tradition or philosophical perspective has the ideal solution to the environmental crisis.* Our approach emphasizes plurality because perhaps we need to recognize that diversity in life forms, in sustained bioregional habitats,

and in cosmological thinking is necessary. Furthermore, it is clear
that theories and practices vary greatly when they are studied in
historical context. In other words, Taoist wisdom regarding nature
was not able to prevent the deforestation of many parts of China,
nor was Buddhist sensitivity to the interconnection of all reality
able to prevent destruction of natural resources in Southeast Asia.
The disjuncture between ideals and reality will temper our expectations; at the same time it will prod us toward more functional
solutions.

It is also important to keep in mind that the traditional world
religions were not faced with the scale of the environmental crisis
of our contemporary world, reeling from the assaults of several centuries of industrialization. Nonetheless, it is significant to draw on
the resources of religion as a means of finding more adequate cosmological perspectives on nature as well as for developing more
functional environmental ethics. It is our contention that significant
changes in attitudes regarding nature will emerge from creating a
comprehensive ethical basis for respecting and preserving nature.
All the major world religions have traditionally developed ethics for
human relations and social interaction. But as Thomas Berry has
shown, while we have moral teachings for homicide and suicide,
they are lacking for biocide or geocide. Our ethics have remained
largely anthropocentric and indifferent to the fate of the natural
world. This is changing gradually as we reexamine the nature of
human-earth relations and begin to build the basis for sustainable
life in the future. It is important to note that "sustainable life" may
not correspond to "sustainable development" as this later term is
increasingly used to promote growth economics with little regard
for inherent limitations on ecological systems.

Worldviews and Ecology, then, provides an overview of various traditional religions and contemporary philosophies as resources for
rethinking the relationship of humans to nature. It is increasingly
clear that the environmental crisis is one of vast scale and complexity. It is also evident that the urgency of the problem is being
raised by many individuals in a variety of disciplines. It is the thesis
of the editors of this volume (crystallized in J. Baird Callicott's
essay) that a new global environmental ethics will be needed to
solve some of the critical issues that face us in the late twentieth
century. As many have noted, we will not preserve what we do not
respect.

It is our contention that various religious and cultural worldviews
have helped shape traditional attitudes toward nature. Indeed, as
Larry Rasmussen observes, from a worldview there emerges a

method for action, from a cosmology there arises an ethics. These are inextricably linked. By presenting various worldviews, we hope that a broadened context for a new ecological ethics will be created. Without such a comprehensive context of restraint and respect, the exploitation of nature and its resources will continue unchecked.

In his introduction to this volume, Tu Wei-ming proposes a reexamination of the resources of the world's great spiritual traditions to help us find a way beyond the Enlightenment mentality. In the second section, religions of Native American, Asian, and Mediterranean peoples are explored for the textual, ritual, and experiential evidence they offer in understanding human-earth relationships. A range of positions is evident—from the biocentric position of Native Americans, to the anthropocentric positions of Judaism, Christianity, Islam, and Baha'i, to the more ecocentric positions of Hinduism, Buddhism, and Jainism, as well as the anthropocosmic orientations of Taoism and Confucianism. In each case, the role of the human in relation to the cosmos is reexamined through the lens of traditional religions.

In the third part of the volume, a range of contemporary ecological perspectives is discussed. The comparative survey by Ralph Metzner sets the tone for this section—describing the paradigm shift in perceptions and values that is taking place in many areas in the transition from the industrial to the ecozoic age. Essays by Charlene Spretnak on ecofeminism, David Ray Griffin on process philosophy, and George Sessions on deep ecology testify to this paradigm shift already occurring. The last two pieces, by Thomas Berry and Brian Swimme, point toward the need for a radical rethinking of the story of the universe—its emergence and evolution—and toward a new understanding of our role in this dynamic, unfolding process.

Worldviews and Ecology, then, is but a beginning in charting the territory of traditions that are being rethought and contemporary worldviews that are being articulated for the first time in response to the current environmental crisis. We hope readers will find in it an invitation for further creative commentaries and worldview to emerge.

Foreword

Some years ago, the late scientist-philosopher Rene Dubos urged the world to "think globally" but to "act locally"—an extremely important exhortation vital to the evolution of a truly humane global community and human survival itself. But perhaps, at the time uttered, this wise counsel may have been somewhat premature. Recent developments, however, have indicated that humanity may have reached the point in its evolution where acting locally and even personally can have global effects, as our population explosion will attest, and that the distinction between local and global has become increasingly blurred. At a time when the future viability of the planet hangs in the balance, local/global must be seen as a premise for a new kind of responsibility for the earth. This unity of spheres has been made even more apparent by human adventures—thanks to space exploration, moonwalks, and the extraordinary advances in communications technology which have now enhanced humanity's ability to conceive of the earth as a whole, literally, to "see globally," as well as to "think globally."

Now that we have been able to see it whole, a foundation may have been laid for a new kind of responsibility for the whole. Likewise, opportunities have been opened to us to transcend polarization and fragmentation and the kind of parochial worldviews that have wreaked such havoc on human thoughts and conduct over time. Integrity for the earth has now become the new global focus; fortunately, more and more philosophers and thinkers are seizing the moment to generate worldviews, systems of thought that might encourage a new spirit of wholeness.

Mary Evelyn Tucker and John A. Grim are to be commended for embarking on this new road to wholeness by the series of essays they have assembled in this issue of the *Bucknell Review, Worldviews and Ecology*. Through these texts we will be able to grasp more fully the basic premises of religious thoughts and insights—whether they be Native American, Jewish, Christian, Buddhist, Confucian, or in the emerging ecological worldviews grounded on an evolutionary cosmology.

We at the United Nations welcome these new perspectives and the remarkable insights now being generated by ecologically premised thinking. These can only help humanity move forward in its search for oneness with the earth and all that is.

Surely, this issue will stimulate new dialogues and encourage our best and brightest to apply intellectual and spiritual gifts for new revelations.

NOEL J. BROWN
Director, United Nations Environment Programme
Regional Office for North America

Worldviews
and Ecology

Beyond the Enlightenment Mentality

Tu Wei-ming
Harvard University

IN my essay "Challenges in Contemporary Spirituality," I, as a student of Asian and comparative religion, made the following observation:

> We need an ethic significantly different from the social Darwinian model of self-interest and competitiveness. We must go beyond the mentality that the promise of growth is limitless and the supply of energy is inexhaustible. The destructiveness of "secular humanism" lies not in its secularity but in its anthropocentrism. While the recognition of the spirituality of matter helps us to appreciate human religiosity as a way of living the fullness of life in all its dimensions, the exclusive focus on humanity as the measure of all things or as endowed with the unquestioned authority of dominion over nature relegates the spiritual realm to irrelevance and reduces nature to an object of consumption. The human project has been so impoverished that the answer to "What is man that thou art mindful of him?" is either want or greed. The crisis of modernity is not secularization per se but the inability to experience matter as the embodiment of spirit.[1]

My observation was occasioned by a powerful image: the celestial vision of the earth, the stunningly beautiful blue planet as seen through the eyes of the astronauts. The image presents two significantly different realities. The unprecedented scientific and technological achievement that enables us not only to survey all boundaries of the good earth but even to measure the thickness of the air we breathe is certainly an established fact. Yet, a more compelling actuality is the realization of how precious and precarious this lifeboat of ours is in the midst of the turbulent ocean of galaxies. This realization, heightened by a poetic sensitivity and infused by a religious sense of awe, impels us to recognize as professionals as well as concerned citizens of the world that we ourselves now belong to the category of the endangered species. This poignant recognition is deduced from the obvious fact that we have mercilessly polluted our own habitat.

19

We may gaze at the distant stars, but we are rooted here on earth and have become acutely aware of its vulnerability and increasingly wary of its fragility. The imagined possibility of creating a new habitat for the human community on an unknown planet by massive emigration has lost much of its persuasive power even in science fiction. The practical difficulty of developing alternative sources of energy and the virtual impossibility of inventing radically different forms of life make us realize how unique is our life on earth. As the horizon of our knowledge extends, we learn that there are limits to the speed and quantity of our economic growth, that natural resources are exhaustible, that the deterioration of our environment has disastrous consequences for the human community as a whole, that the serious loss of genes, species, and ecosystems is endangering the equilibrium of our life-support system, and that a minimum condition for continuous human survival requires the actual practice of sustainable life in highly industrialized societies. The painful acknowledgment that what we have been doing to nature in the last two centuries since the French Revolution, especially in the last four decades since the Second World War, has resulted in a course of self-destruction that has instilled in us a sense of urgency. Indeed, by poisoning the air we breathe and the water we drink, in short, by degrading our environment, we are recklessly reducing the livability of our habitat to a point of no return. The necessity of a basic reorientation of our thought with a view toward a fundamental restructuring of our style of life is glaringly clear.

Tools and methods specifically designed to reduce the magnitude of environmental degradation, such as recycling aluminum and applying pollution-control technology, are now readily available. The concern for halting the trend toward massive destruction of biodiversity has prompted new frontiers of research in ecological science. Furthermore, in economics, the emerging field of ecological economics has already recommended ways for using economic manipulations in favor of conservation. However, as Lawrence Hamilton poignantly reminds us, "they do not get at the cause of the problems."[2] Far-sighted ecologists, engineers, economists, and earth scientists, intent on developing a communal critical self-consciousness for "saving spaceship earth," have made an appeal to poets, priests, artists, and philosophers for their active participation in this intellectual and spiritual joint venture to make our habitat, our home, safe for generations to come. The felt need to focus our attention on ethics, values, and religions as ways of "caring for the planet and reducing its rate of impoverishment" is urgent. As we diagnostically and prognostically address issues pertaining to conserving biological diversity, the cooperation of scholars in the natural sciences, social sciences, and humanities is necessary. In this essay, I would like to discuss the Enlightenment mentality so that we fully acknowledge a persistent psychocultural cause underlying the destructive power of these "transnational, transgenerational, and transideological" assaults on the environment. I hope to bring some understanding to a major paradox as we reflect upon our human condition in a way scientifically disinterested and yet profoundly personal.

The Enlightenment mentality underlies the rise of the modern West as the most dynamic and transformative ideology in human history. Virtually all major spheres of interest characteristic of the modern age are indebted to or intertwined with this mentality: science and technology, industrial capitalism, market economy, democratic polity, mass communication, research universities, civil and military bureaucracies, and professional organizations. Furthermore, the values we cherish as definitions of modern consciousness, including liberty, equality, human rights, the dignity of the individual, respect for privacy, government for, by and of the people, and due process of law are genetically, if not structurally, inseparable from the Enlightenment mentality. We have flourished in the spheres of interest and their attendant values occasioned by the advent of the modern West since the Enlightenment of the eighteenth century. They have made our lifeworld operative and meaningful.

We are so seasoned in the Enlightenment mentality that we assume that the reasonableness of its general ideological thrust is self-evident. The Enlightenment faith in progress, reason, and individualism may have lost some of its persuasive power in the modern West, but it remains a standard of inspiration for intellectual and spiritual leaders throughout the world. It is inconceivable that any modern project, including those in ecological sciences, does not subscribe to the theses that the human condition is improvable, that it is desirable to find rational means to solve the world's problems, and that the dignity of each person as an individual ought to be respected. Enlightenment as human awakening, as the discovery of the human potential for global transformation, and as the realization of the human desire to become the measure and master of all things is still the most influential moral discourse in the political culture of the modern age; for decades it has been the unquestioned assumption of the ruling minorities and cultural elites of the developing countries, as well as the highly industrialized nations.

A fair understanding of the Enlightenment mentality requires a frank discussion of the dark side of the modern West as well. The "unbound Prometheus," symbolizing the runaway technology of development, may have been a spectacular achievement of human ingenuity in the early phases of industrial revolution. Despite impassioned reactions from the Romantic movement and insightful criticisms of the forefathers of the "human sciences," the Enlightenment mentality fueled by the Faustian drive to explore, to know, to conquer, and to subdue persisted as the reigning ideology of the modern West.

By the late nineteenth century, the Enlightenment mentality, revealing itself as "knowledge is power" (Francis Bacon), the historical inevitability of human progress (August Comte), or "the humanization of nature" (Karl Marx), had become an intellectual source for social Darwinian competitiveness. This competitive spirit, justified by a simpleminded reading of the principle of "the survival of the fittest," in turn provided a strong rationale for imperialism. To be sure, according to Max Weber, the rise of the modern West owes much to the Protestant work ethic which historically engendered the spirit of capitalism in Western Europe and North America. Nevertheless, modernization, as rationalization, is Enlightenment mentality to the core. Faith in progress, reason, and individualism propelled the modern West to engulf the world in a restless march toward modernity. As the Western nations assumed the role of innovators, executors, and judges of the international rules of the game defined in terms of competition for wealth and power, the stage was set for growth, development, and exploitation. The unleashed juggernaut blatantly exhibited unbridled aggressiveness toward humanity, nature, and itself. This unprecedented destructive engine has for the first time in history made the viability of the human species problematical.

The realization that the human species may not be viable and that human life as lived in the last two centuries has explosive potential for destroying the entire life-support system has prompted some reflective and concerned minds in the natural sciences, social sciences, and humanities to join forces in a concerted effort to think through the issue in the broadest terms possible and to act immediately and concretely in order to bring about realizable incremental results.

The spirit of "thinking globally and acting locally" enables us to put the *Problematik* at hand in proper perspective. Values espoused by the French Revolution, namely liberty, equality, and fraternity, as well as the aforementioned progress, reason, and individualism embedded in the Enlightenment mentality, are integral aspects of our heritage. We do well to recognize the persuasiveness of these values throughout the world and to affirm our commitment to them for giving meaning to our cherished way of life. The lamentable situation that these values are being realized only in Western Europe and North America must not be used as an excuse to relegate them to a culturally specific and thus parochial status. Notwithstanding the tremendous difficulty of spreading these values to other parts of the world, the potential for their universalizability is widely acknowledged. The most formidable defenders of these values are not necessarily found in Paris, London, or New York; they are more likely to be found in Beijing, Moscow, or New Delhi.

A brief look at what Talcott Parsons defined as the three inseparable dimensions of modernity two decades ago will help to sharpen our focus on the issue. Despite the acknowledgment that it has taken centuries for democracy to flourish in England, France, or the United States and that the forms it has taken in these societies are still seriously flawed, democracy as a standard of inspiration has universal appeal. Moreover, the "third wave of democracy" is a major transformative force in international politics. A more powerful dynamic can be seen working in the competitive markets. The disintegration of communist Eastern Europe and the collapse of the Union of Soviet Socialist Republics clearly indicate the strength of democratic polity and market economy in defining the process of modernization. Although, individualism, Parsons's third dimension of modernity, is less persuasive, it seems to symbolize an ethos underlying the entire value system of the modern West.[3]

While we are willing to grant that the modernization project as exemplified by the modern West is now the common heritage of humanity, we should not be blind to the serious contradictions inherent in the project and the explosive destructiveness embodied in the dynamics of the modern West. The legacy of the Enlightenment is pregnant with disorienting ambiguities. The values it espouses do "not cohere as an integrated value system recommending a coordinated ethical course of action."[4] The conflict between liberty and equality is often unresolvable. It may not be farfetched to suggest, in grossly simplified terms, that while capitalist countries have embraced principles of liberty to organize their political life, communist societies have articulated the rhetoric of equality to impose their ideological control. The matter is greatly complicated by the deliberate attempts of the capitalist countries to employ socialist measures, ostensibly to blunt the hard edges of free enterprise but, in reality, to save capitalism from collapsing since the end of the First World War.

Classical liberalism, as brilliantly developed by Friedrich von Hayek, has performed an invaluable service to elucidate the dangers of socialism as a "road to serfdom," but its own role and function in providing both theoretical and practical guidance to advanced capitalism are quite limited. The idea of competitive market or free enterprise, in Adam Smith's sense, may have been both a motivational force and an ideological weapon in the modernizing process, but it has never been fully implemented as a political or economic institution. In fact, the enormous growth of the central government, not to mention the ubiquity of the military bureaucracy, in all Western democracies has so fundamentally redefined the insights of the Enlightenment that self-interest, expansion, domination, manipulation, and control have supplanted seemingly innocuous values such as progress, reason, and individualism. A realistic appraisal of the Enlightenment mentality reveals many faces of the modern West to be incongruous with the image of "the Age of Reason." In the context of modern Western hegemonic discourse, progress means inequality, reason means self-interest, and individualism means greed. The American dream of owning a car and a house, earning a fair wage, and enjoying freedom of privacy, expression, religion, and travel, while reasonable to our sense of what ordinary life entails, is lamentably unexportable as a modern demand from a global perspective.

An urgent task for the community of like-minded persons deeply concerned about ecological issues is to insure that both the ruling minorities and cultural elites in the modern West actively participate in this spiritual joint venture to rethink the Enlightenment heritage. The paradox is that we cannot afford to uncritically accept its inner logic in light of the unintended negative consequences it has engendered for the life-support system; nor can we reject its relevance, with all of the fruitful ambiguities it entails, to our intellectual self-definition, present and future. There is no easy way out. We do not have an "either-or" choice. The possibility of a radically different ethic or a new value system separate from and independent of the Enlightenment mentality is neither realistic nor authentic. It may even appear to be either cynical or hypercritical. We need to explore the spiritual resources that may help us to broaden the scope of the Enlightenment project, deepen its moral sensitivity, and, if necessary, creatively transform its genetic constraints in order to fully realize its potential as a worldview for the human community as a whole.

A key to the success of this spiritual joint venture is to recognize the conspicuous absence of the idea of community, let alone the global community, in the Enlightenment project. Fraternity, a functional equivalent of community in the three cardinal virtues of the French Revolution, has received scanty attention in modern Western economic, political, and social thought. The willingness to tolerate inequality, the faith in the salvific power of self-interest, and the unbridled affirmation of aggressive egoism have greatly poisoned the good will of progress, reason, and individualism. The first step in creating a new world order is to articulate a universal intent for the formation of a global community. This requires, at a minimum, the replacement of the principle of self-interest, no matter how broadly defined, with a new golden rule: "Do not do unto others what you would not want others to do unto you." Since the new golden rule is stated in the negative, it will have to be augmented by a positive principle: "in order to establish myself, I must help others to establish themselves; in order to enlarge myself, I have to help others to enlarge themselves." An inclusive sense of community, based on the communal critical self-consciousness of the reflective and concerned ecological minds, may emerge as a result.

The mobilization of three kinds of spiritual resources is necessary to insure that this simple vision be grounded in the historicity of the cultural complexes informing our ways of life today. The first kind involves the ethicoreligious traditions of the modern West, notably Greek philosophy, Judaism, and Christianity. The very fact that they have been instrumental in giving birth to the Enlightenment mentality makes a compelling case that they reexamine their relationships to the rise of the modern West in order to create a new public sphere for the transvaluation of typical Western values. The dichotomizing of matter/spirit, body/mind, sacred/profane, man/nature, or creator/creature must be transcended to allow supreme values such as the sanctity of the earth, the continuity of being, the beneficiary interaction between the human community and nature, and the mutuality between humankind and heaven to receive the saliency they deserve in philosophy and theology.

The Greek philosophical emphasis on rationality, the biblical image of man having "dominion over the fish of the sea, and over the fowl of the air, and over every living thing that moveth upon the earth," and the so-called Protestant work ethic provided necessary, if not sufficient, sources for the Enlightenment mentality. However, the unintended negative consequences of the rise of the modern West have so undermined the sense of community implicit in the Hellenistic idea of the citizen, the Judaic idea of the covenant, and the Christian idea of universal love that it is morally imperative for these great traditions, which have maintained highly complex and tension-ridden relationships with the Enlightenment mentality, to formulate their critique of the blatant anthropocentrism inherent in the Enlightenment project.

The second kind of spiritual resources are derived from non-Western axial-age civilizations which include Hinduism, Jainism, and Buddhism in South and Southeast Asia, Confucianism and Taoism in East Asia, and Islam. These ethicoreligious traditions provide sophisticated and practicable resources in worldviews, rituals, institutions, styles of education, and patterns of human relatedness. They can help to develop styles of life, both as continuation of and alternative to the Western European and North American exemplification of the Enlightenment mentality. Industrial East Asia, under the influence of Confucian culture, has already developed a less adversarial, less individualistic, and less self-interested modern civilization. The coexistence of market economy with government leadership, democratic polity with meritocracy, and individual initiatives with group orientation has made this region economically and politically the most dynamic area of the world since the Second World War. The implications of the contribution of Confucian ethics to the rise of industrial East Asia for the possible emergence of Hindu, Jain, Buddhist, and Islamic forms of modernity are far-reaching. The westernization of Confucian Asia (including Japan, the two Koreas, mainland China, Hong Kong, Taiwan, Singapore, and Vietnam) may have forever altered its spiritual landscape, but its indigenous resources (including Mahayana Buddhism, Taoism, Shintoism, shamanism, and other folk religions) have the resiliency to resurface and make their presence known in a new synthesis. The caveat, of course, is that, having been humiliated and frustrated by the imperialist and colonial domination of the modern West for more than a century, the rise of industrial East Asia symbolizes the instrumental rationality of the Enlightenment heritage with a vengeance. Indeed, the mentality of Japan and the Four Mini-Dragons is characterized by mercantilism, commercialism, and international competitiveness. Surely, the possibility of their developing a more humane and sustainable community should not be exaggerated—nor should it be undermined.

The third kind of spiritual resources involve the primal traditions: Native American, Hawaiian, Maori, and numerous tribal indigenous religious traditions. They have demonstrated with physical strength and aesthetic elegance that human life has been sustainable since the Neolithic age. The implications for practical living are far-reaching. Their style of human flourishing is not a figment of the mind but an experienced reality in our modern age.

A distinctive feature of primal traditions is a profound sense and experience of rootedness. Each indigenous religious tradition is embedded in a concrete place symbolizing a way of perceiving, a mode of thinking, a way of living, an attitude, and a worldview. Can we learn from Native Americans, Hawaiians, and others whom we often refer to as "primal" peoples? Can they help us solve our ecological crisis?

Given the unintended disastrous consequences of the Enlightenment mentality, there are obvious lessons that the modern mindset can learn from indigenous religious traditions of primal peoples. A natural outcome of primal peoples' embeddedness in concrete locality is their intimate and detailed knowledge of their environment; indeed the demarcations between their human habitat and nature are muted. Implicit in this model of existence is the realization that mutuality and reciprocity between the anthropological world and the cosmos at large is both necessary and desirable. What we can learn from them, then, is a fundamental restructuring of our way of perceiving, thinking, and living; we are urgently in need of a new attitude and a new worldview. A critique of the Enlightenment mentality and its derivative modern mindset from primal consciousness as interpreted by the concerned and reflective citizens of the world could be thought provoking.

An equally significant aspect of the primal way of living is the ritual of bonding in ordinary daily human interaction. The density of kinship relations, the rich texture of interpersonal communication, the detailed and nuanced appreciation of the surrounding natural and cultural world, and the experienced connectedness with ancestors point to communities grounded in ethnicity, gender, language, land, and faith. The primordial ties are constitutive parts of their being and activity. In Huston Smith's characterization, what they exemplify is participation rather than control in motivation, empathic understanding rather than empiricist apprehension in epistemology, respect for the transcendent rather than domination over nature in worldview, and fulfillment rather than alienation in human experience.[5] As we begin to question the soundness or even sanity of some of our most cherished ways of thinking—such as regarding knowledge as power rather than wisdom, asserting the desirability of material progress despite its corrosive influence on our soul, and justifying the anthropocentric manipulation of nature even at the cost of destroying the life-support system—primal consciousness emerges as a source of inspiration.

A scholar of world spirituality, Ewert Cousins, in response to the ecological crisis, poignantly remarks that, as we look toward the twenty-first century with all the ambiguities and perplexities we experience, earth is our prophet and the indigenous peoples are our teachers.[6] Realistically, however, those of us who are seasoned in the Enlightenment mentality cannot abdicate the hermeneutic responsibility to interpret the meaning of the earth's prophecy and to bring understanding to the primal peoples' message. The challenge is immense. For the prophecy and the message to be truly heard in the modern West, they may have to be mediated through dialogue with non-Western axial-age civilizations. This combined effort is necessary to enable primal consciousness to be fully present in our self-reflexivity as we address issues of globalization.

I am proposing that, as both beneficiaries and victims of the Enlightenment mentality, we show our fidelity to our common heritage by enriching it, transforming it, and restructuring it with all three kinds of spiritual resources still available to us for the sake of developing a truly ecumenical sense of global community. Our approaches, while divergent in their methodologies and different in their ethical and religious orientations, are all serious attempts to identify and tap the spiritual resources available in the human community for inspirational guides to find a way out of our predicament: the road to liberation may mislead us to the dark cave of an "endangered species." It may not be immodest to say that we are beginning to develop a fourth kind of spiritual resources from the core of the Enlightenment project itself. Our disciplined reflection, a communal act rather than an isolated struggle, is a first step toward the "creative zone" envisioned by religious leaders and ethical teachers.

Notes

This essay, originally written as an epilogue for a soon-to-be-published collection of papers on "Relating Ethics, Culture, and Religion to the Conservation of Biological Diver-

sity," edited by Lawrence S. Hamilton, was presented for discussion at the Fourth Conference on World Spirituality, sponsored by the Dialogue of Civilizations Project at the East-
West Center in Honolulu in June 1992.

1. In *Local Knowledge, Ancient Wisdom*, ed. Steven Friesen (Honolulu: East-West Center,
1991), 2–3.

2. I am grateful to Dr. Lawrence S. Hamilton of the Environment and Policy Institute
at the East-West Center. His commitment to bringing dimensions of human behavior and
thought into the scientific discussion of biological diversity and his insistence that without
the active participation of the humanists "the best attempts of natural scientists are destined to failure" was a source of inspiration for my reflection on the "Enlightenment
Mentality." The unidentified quotations in this essay are from his introduction to the
projected volume referred to in the headnote above. I would also like to note that similar
lines of thinking are further explored in my "Core Values and the Possibility of a Fiduciary
Global Community" in *Restructuring for World Peace on the Threshold of the Twenty-First Century*, ed. Katharine Tehranian and Majid Tehranian (Cresskill, N.J.: Hampton Press,
1992), 333–45.

3. See Talcott Parsons, *The System of Modern Societies* (Englewood Cliffs, N.J.: Prentice-
Hall, 1971), 114ff.

4. Tu Wei-ming, "Intellectual Effervescence in China," *Daedalus* 121 (Spring 1992): 257.

5. Huston Smith, *The World's Religions* (New York: HarperCollins, 1991), 365–83.

6. See Ewert H. Cousins, "Three Symbols for the Second Axial Period," in Friesen, ed.,
Local Knowledge, Ancient Wisdom. Also, see Ewert H. Cousins, *Christ of the 21st Century*
(Rockport, Mass.: Element, 1992), 105–31.

Toward a Global Environmental Ethic

J. Baird Callicott

University of Wisconsin, Stevens Point

ENVIRONMENTAL ethics, as a distinct subject of disciplined inquiry, came into being in the early 1970s in response to the sudden recognition in the 1960s that industrial civilization had engendered an environmental crisis. However, vernacular environmental ethics have existed implicitly in many indigenous and traditional cultures throughout the world for a very long time, and in the West, protoenvironmental ethics were adumbrated by Henry David Thoreau, John Muir, and Aldo Leopold.

When first recognized, the environmental crisis, though very widespread, was nevertheless perceived to be an aggregation of local phenomenon—an oil spill off the shore of Santa Barbara, California, mercury pollution in Japan's Miamata Bay, acid precipitation in Eastern Europe, an epidemic of schistosomiasis behind the Aswan dam in Egypt, etc. In response, several approaches to formal environmental ethics have been articulated over the past quarter century.

One approach, now called "anthropocentrism" (about which more later), simply applied standard Western moral philosophies—utilitarianism, for example—to such novel problems. The theories were familiar and tried, if not true; only the casuistry was new.

A second approach, now called "biocentrism" (literally, life centered) built on the simultaneously developing work of the animal welfare ethicists and attempted to extend familiar human-to-human ethics beyond sentient animals to all living things. To provide a rationale for lowering the qualification for ethical standing is the central strategy of this approach. Classically, a being had to be rational to qualify for moral consideration. But if that criterion were taken seriously, the animal welfare ethicists argued, many nonrational human beings (infants, the severely retarded, and the senile) would be morally disenfranchised and exposed to all the horrors at human hands to which many animals are routinely subjected. To cover such human beings, we must lower the criterion of moral considerability—from reason to sentience—covering at the same time many kinds of animals. But why stop at sentience? Nonsentient animals and plants also have goods of their own, whether they care about them or not; they too may be benefited or harmed, even if they may not know it; they have interests, despite the fact that they are not interested in their interests. We ought to give due consideration to the interests of all living things, argued the biocentrists.

Perhaps, but our environmental problems—the very problems that provoked a search for a new environmental ethic—have little to do with the welfare of individual shrubs, bugs, and grubs. They have a more holistic cast. Environmentalists are concerned about the disappearance of species, about the degradation of ecosystems, about water pollution and soil erosion. A third, more holistic approach, called "ecocentrism," built upon the classic land ethic of Aldo Leopold. Following Darwin, Leopold believed that ethics originated as a means to social organization and that we human beings have duties and obligations to the several communities to which we belong, as well as to individual fellow members. Ecology, he simply pointed out, has recently discovered that each of us is a member of a local biotic community, as well as of various human communities. Thus he argued that a "land ethic changes the role of Homo sapiens from conqueror of the land community to plain member and citizen of it." And further that "it implies respect for fellow-members, and also respect for the community as such." Summarizing the land ethic, Leopold averred that "a thing is right when it tends to preserve the integrity, stability, and beauty of the biotic community; it is wrong when it tends otherwise."[1]

Now, however, in addition to local events, such as those that recently occurred in Prince William Sound, Bhopal, and Chernobyl, the environmental crisis is understood to have an additional, global dimension—global warming, stratospheric ozone thinning, abrupt massive species extinction, destruction of the world's girdle of tropical forests, etc. Reflecting the changed bilevel understanding of the contemporary environmental crisis—it manifests itself on both local and global scales—we shall need to complement the variety of diverse local environmental ethics with a univocal global environmental ethic. Here I suggest how a globalized ecocentric environmental ethic, like the Leopold land ethic, based upon contemporary ecology might serve as a common international environmental ethic and how it might interface with the local environmental ethics implicit in the world's many indigenous and traditional cultures.

Though the people of the earth are all members of one species and share one ecologically integrated planet, we live, nevertheless, in many and diverse worlds. Each contemporary society, moreover, at once lives in a planetary culture—united by economic interdependency, jet transport, and satellite communication systems—*and* in a separate reality shaped by its formerly isolated cognitive cultural heritage. The revival and deliberate construction of environmental ethics from the raw materials of indigenous, traditional, and contemporary cognitive cultures represent an important and essential first step in the future movement of human material cultures toward a more symbiotic relationship, however incomplete and imperfect, with the natural environment. The effort to mutually tune the resulting diverse environmental ethics—to achieve some orchestration of the chorus of voices singing of a human harmony with nature—represents an important and essential second step toward the same desideratum.

Until recently, it may seem, human material culture, human technology, was impotent to affect seriously the natural environment for better or worse. Since preindustrial Homo sapiens apparently posed no serious threat to the natural environment, it would seem that indigenous and traditional environmental ethics would not exist—because they would have been otiose and unnecessary.

However, a reexamination of human history and prehistory from an ecological perspective reveals a long-standing, but changing pattern of anthropogenic environmental degradation. Paleolithic hunter-gatherers armed with stone-tipped spears and arrows, snares and traps, and (not least) fire may have caused local extirpation and, in some cases, may have played a role in the total extinction of other animal species. Certainly, in any case, prehistoric Homo sapiens profoundly altered the character of biotic communities. Neolithic, ancient, medieval, and modern agriculturalists caused soil erosion, siltation of surface waters, deforestation, salinization of both arable lands and fresh waters, and desertification.

A reexamination of human history and prehistory also reveals the existence of culturally evolved and integrated environmental ethics which served to limit the environmental impact of preindustrial human technologies. In many indigenous cultures nature was represented as enspirited or divine (the ancient Japanese *kami* are a leading example)—and therefore the direct object of respect or of reverence; in some traditional cultures (among the Hebrews, for example, in the Middle East) nature was the creation of God— and thus should be used with care and passed on intact; in others (ancient Chinese Taoism is a leading example), man was thought to be part of nature—and a good human life was understood therefore to be one in harmony with nature; in still others, a oneness of all life was envisioned (called *Brahman* in Advaita Vedanta)—together with an attitude of *ahimsa* in respect to all living things; and so on. In *Nature in Asian Traditions of Thought,* a collection of essays that I edited with Roger T. Ames, the conceptual resources for environmental ethics in Hinduism, Buddhism, Taoism, and other traditions of thought are explored and developed by a number of distinguished scholars. I have expanded this process of intellectual recovery in a more recent work entitled *The World's Great Ecological Insights.*[2]

Industrial civilization, of course, has polluted the environment with synthetic toxic chemicals and radioactive elements, as well as intensified the kinds of environmental mischief already afoot in the activities of preindustrial people. With the emergence of an industrial human culture of global reach, the human impact on nature has so increased in force, intensity, and ubiquity that under the worst possible scenario imaginable, thermonuclear holocaust, people may utterly destroy the biosphere (at least as we now know it) along with ourselves. Short of this cataclysmic and apocalyptic event, the global ecosystem may gradually be degraded to the extent that many higher forms of life presently existing (including Homo sapiens) will no longer be adapted to its radically changed conditions.

The emergence of global industrial human culture was ironically accompanied by a loss of the sorts of preindustrial environmental ethics just mentioned. The secularism, humanism, and materialism of industrial culture demystified and undermined earlier environmental ethics, aggravating the destructive impact of industrial technology. Here this is an irony: just when we need environmental ethics more than ever, global industrial civilization with its infininitely greater power for environmental destruction eclipsed the environmental ethics (along with many other traditional cultural values) that prevailed in the past and that served to restrain traditional human patterns of resource exploitation.

The secularism, humanism, and materialism characteristic of contemporary industrial culture has, on the other hand, evolved a protean social ethic peculiar to itself. The moral concept lying at its core is the intrinsic value, autonomy, and dignity of individual human beings (as glossed by Descartes, Hobbes, Locke, and other early modern philosophers). Two complementary (and just as often competing) streams of moral philosophy have flowed from this central source: utilitarianism (first set out by Jeremy Bentham), which emphasizes human welfare, and deontology (first set out by Immanuel Kant), which emphasizes human dignity as a basis for human rights.

A latter-day modern secular environmental ethic may be developed as an addendum to the moral implications devolving from consideration of human welfare and human rights. With the emergence of the science of ecology and related sciences, it is now painfully clear that human actions which have direct deleterious effects on the environment often also have indirect deleterious effects on human beings. For example, cutting and burning a moist tropical forest in order to create pasture directly destroys an ecosystem and its nonhuman native denizens, but it may also indirectly adversely affect aggregate human welfare because of the now well-understood ecological, climatological, hydrological, and erosional effects of deforestation—to say nothing of its adverse impact on the indigenous peoples subsisting there.

From a classical utilitarian point of view, massive tropical deforestation would appear to be unethical because it benefits a few people (lumber and cattle barons) in the short run at the expense of many people (indigenous forest dwellers, the local landless population, and, less directly, people everywhere in the world) now and in perpetuity. From a classical human rights point of view, the immorality of deforestation is less clear because of the historical conflation of human rights generally with human property rights more particularly and with "free enterprise." But again, the exercise of one person's rights, in theory, are limited by those of others. And, if my moral sensibilities are not in error, increasingly these days, "human rights" are construed more broadly to include, in addition to the right to political liberties and unfettered economic activity, the right to certain amenities— subsistence with dignity, access to rudimentary education and basic health care, and a viable, livable natural environment.

One might thus go on to develop a modern secular environmental ethic erected upon the twin pillars of human welfare and human rights. Environmental ethics would thus consist of a thorough integration of environmental science and technical expertise with the conventional values of contemporary industrial civilization. A contribution to a mature environmental ethics, so conceived, would attempt to predict the effects on human welfare and human rights (broadly construed) of human behaviors that have environmental impact. State-of-the-art utilitarian and rights theory is only a little less complicated and sophisticated than state-of-the-art environmental impact assessment. By combining the two—no small task—human environmental behavior could be ethically evaluated.

The biological sciences, especially ecology and the theory of evolution, in tandem with the theories of special and general relativity and quantum theory (together sometimes called the "new physics") are creating a new postmodern scientific worldview. In it, human beings are thoroughly embedded in nature: kin to all other living things on planet Earth and systemically interrelated with them. There is another, stronger, more direct, approach to environmental ethics which is more resonant with this emerging new scientific worldview. Such an approach to environmental ethics would make the effects of human actions upon individual nonhuman natural entities and nature as a whole directly accountable—irrespective of their indirect effects upon human welfare. Such an environmental ethic would be stronger, as well as more direct, since it could ethically assay environmentally destructive human action which had little or no negative effect on human beings. And it has another appeal. It is more consonant with the environmental ethics of preindustrial cultures. An environmental ethic based upon human welfare and rights would show a decided Western bias in its strident anthropocentrism and individualism, attitudes not characteristic of most other cultural worldviews.

An environmental ethic that takes into account the impact of human actions directly upon nonhuman natural entities and nature as a whole is called an ecocentric environmental ethic. An ecocentric environmental ethic is supported by the evolutionary, ecological, foundational, and cosmological dimensions of the presently evolving postmodern scientific worldview. Since most indigenous and traditional environmental ethics also fit the ecocentric mold, we can envision a solidarity of an international ecocentric environmental ethic with the many and diverse traditional environmental ethics.

With the current and more ominous global dimension of the twentieth century's environmental crisis now at the forefront of attention, environmental philosophy must strive to facilitate the emergence of a global environmental consciousness that spans national and cultural boundaries. In part, this requires a more sophisticated cross-cultural comparison of traditional and contemporary concepts of the nature of nature, human nature, and the relationship between people and nature than has so far characterized discussion.

Western philosophers looked initially to traditional Eastern wisdom in their search, begun in earnest in the late 1960s, for an environmental ethic located in a deep ecological consciousness. And in fact Eastern philosophy has historically shaped the gradually emerging environmental consciousness in the West. The American Transcendentalism of Ralph Waldo Emerson and Henry David Thoreau—who were among the first American thinkers to look upon nature as something more than an obstacle to progress and a pool of natural resources—was inspired by Hindu thought. Further, Arne Naess was inspired by the Vedantic doctrine of nonduality to make cultivation of the experience of oneness with nature the core practice of deep ecology.

In the mid–twentieth century the emerging contemporary environmental movement was profoundly influenced by Japanese Zen Buddhism. Zen had been powerfully and persuasively represented in the West by D. T. Suzuki in the early twentieth century. Alan Watts popularized Suzuki's somewhat more academic representation. The American nature poet, Gary Snyder, inspired by Watts, studied Zen Buddhism in Kyoto. In his eventual work, a raw and uncultivated American love of and sensitivity to nature was integrated with the very advanced natural aesthetic cultivated for centuries in Japan. Snyder was a charter member of the mid-century American counterculture which called itself the Beat Generation—romanticized by the enormously popular novelist, Jack Kerouac, in the book *Dharma Bums*. Thus when Americans awakened to the environmental crisis in the late 1960s, they turned for philosophical guidance to the cultural alternatives then popular, and Zen Buddhism was by far the most in evidence. Since then, the attention of Western environmental philosophers has gravitated more to Taoism. The concept of living in accordance with the *tao* of nature complements the evolutionary and ecological axiom that human beings are part of nature and must conform human ways of living to natural processes and cycles. Especially in the Taoist concept of *wu wei*, Western environmental ethicists have found a traditional Eastern analogue of what they call appropriate technologies—technologies that blend with and harness natural forces as opposed to technologies that resist and attempt to dominate and reorganize nature.

I am convinced that the intellectual foundations of the industrial epoch in world history are an aberration. I agree with Fritjof

Capra that a new paradigm is emerging that will sooner or later replace the obsolete mechanical worldview and its associated values and technological esprit. I agree further with Capra, as he broached the idea in both the *Tao of Physics* and *The Turning Point*, that the emerging twenty-first century paradigm has many conceptual affinities with preindustrial natural attitudes and values, especially those of the East.[3] Thus, detailed cross-cultural comparison of traditional concepts of the nature of nature, human nature, and the relationship between people and nature with the ideas emerging in ecology and the new physics should be mutually reinforcing. On the one hand, traditional environmental ethics can be thus revived and, just as importantly, validated or verified by the affinity of their foundational ideas with the most exciting new ideas in contemporary science. On the other hand, the otherwise abstract and arcane concepts of nature, human nature, and the relationship between people and nature implied in ecology and the new physics can be expressed and articulated in the rich vocabulary of metaphor, simile, and analogy developed in the traditional sacred and philosophical literature of the world's many and diverse cultures.

What I envision for the twenty-first century is the emergence of an international environmental ethic based upon the theory of evolution, ecology, and the new physics and expressed in the cognitive lingua franca of contemporary science. Complementing such an international, scientifically grounded and expressed environmental ethic—global in scope as well as focus—I also envision the revival of a multiplicity of traditional cultural environmental ethics that resonate with it and that help to articulate it. Thus we may have one worldview and one associated environmental ethic corresponding to the contemporary reality that we inhabit one planet, that we are one species, and that our deepening environmental crisis is common and global. And we may also have a plurality of revived and renewed traditional worldviews and associated environmental ethics corresponding to the historical reality that we are many peoples inhabiting many diverse bioregions apprehended through many and diverse cultural lenses. But this one and these many are not at odds. Quite the contrary, they may be regarded as a single but general and abstract metaphysic and moral philosophy expressible in many conceptual modes. Each of the many worldviews and associated environmental ethics may crystalize the international ecological environmental ethic in the vernacular of a particular and local cultural tradition. The environmental ethic based upon contemporary international science and those implicit in the many indigenous and traditional cultures can thus be fused to form a unified but multifaceted global environmental ethic. Let us by all means think globally and act locally. But let us also think locally as well as globally and try to tune our global and local thinking as the several notes of a single, yet common, chord.

Notes

1. Aldo Leopold, *A Sand County Almanac and Sketches Here and There* (New York: Oxford University Press, 1949).

2. J. Baird Callicott and Roger T. Ames, eds., *Nature in Asian Traditions of Thought: Essays in Environmental Philosophy* (Albany: State University of New York Press, 1989), and J. Baird Callicott, *The World's Great Ecological Insights: A Critical Survey of Traditional Environmental Ethics from the Mediterranean Basin to the Australian Outback* (Berkeley: University of California Press, 1994).

3. Fritjof Capra, *The Tao of Physics: An Exploration of the Parallels between Modern Physics and Eastern Mysticism* (Boston: Shambhala, 1975), and Fritjof Capra, *The Turning Point: Science, Society, and the Rising Culture* (New York: Simon & Schuster, 1982).

Traditional World Religions

Native North American Worldviews and Ecology

John A. Grim

Bucknell University

A NY discussion of Native American worldviews and ecology
should begin with an appreciation for the diverse relationships between various human communities and the land on which
particular traditions have been nurtured. From the pre-Colum-
bian era to the present there have been not one but many Native
North American worldviews and religious practices. Native
spokespeople say that there are over five hundred distinct cultural
traditions which still maintain sacred relations with the diverse
terrains and life forms of the North American continent.[1] Thus,
there are a variety of indigenous positions on ecological issues.
Moreover, there are differences *between* native peoples as well as
within any one native community.

Acknowledging the diverse voices of individual native spokespeople as well as the differences of indigenous groups enables
us to explore contemporary concerns and insights among Native
American communities regarding the bioregions in which they
live. By "indigenous peoples of North America" I am referring
to the ethnic groups who are variously called American Indians,
First Peoples, and Native Americans. In all instances I will endeavor to use the indigenous name of the people being discussed
but I will also use the alternative terms as a reminder to the reader
of the extreme diversity of indigenous peoples in North America.

The question of difference also emphasizes the gap between
indigenous and dominant societies. The value placed on sacred
relationships with one's homeland among indigenous North
American peoples is not simply a nationalistic exploitation which
can be folded into patriotic symbols of one's country. Nor is the
ecorelationship established by indigenous peoples with their bioregion presented here as an ideological position to be cloned by
dominant America. Rather, the ecological reciprocity presented
in the religions of indigenous peoples raises questions regarding

a functional cosmology, environmental ethics, and appropriate rituals which such a cosmology evokes to sustain human-earth relations. Responses to these questions are made by indigenous peoples in terms of their particular worldviews and with an understanding that these responses are interrelated. Referring to the lived experience of these interrelated responses as a "lifeway" is an intentional emphasis in this essay on both the pragmatic and motivational aspects of a functional cosmology.[2] That is, a story of the world which informs all aspects of life among a people, giving subsistence practices, artistic creation, ritual play, and military endeavor a significant context often discussed as "religion."

"Lifeways" refers to this functional interaction of cosmology and cultural activity. The spiritual dimensions of traditional Native North American cultures were established in sociohistorical contexts inseparable from their modes of subsistence, their social forms of organization, and their views of the world. Historical encounters with now-dominant societies have certainly changed these indigenous traditions. Continuity with traditional precontact spiritual wisdom is a question which can be raised in an analytical context without undermining the integrity, coherence, or meaning of contemporary indigenous lifeways. These analytical questions, however, are quite different from the synthetic modes of thought active among indigenous peoples which seek to interrelate cosmology and particular cultural activities.

Prominent among the critiques coming from both disaffected and secularized indigenous Americans as well as from critical mainstream Americans is that of the authenticity of Native North American worldviews. After five hundred years of fragmenting contact following invasion, colonial settlement, and modern urbanization, what is left of the lifeway traditions of the First Peoples? Moreover, how can contemporary non-Native Americans, who have lost touch so long ago with affective relations to a living, personified land, even begin to understand the significance of the Native American attitudes toward local environments. The Lakota lawyer and intellectual, Vine Deloria, Jr., reflected on these issues when he wrote:

> When the old circles or hoops of life were broken, thousands of years ago for most non-Indians and a century ago for most Indians, the possibility of recapturing that original sense of awe and respect [for the earth] was lost and could not be recovered. We have simply been playing out the logical possibilities of what the fragments of those original hoops made available to people. The remaining ceremonies

and traditional practitioners may well serve as focal points around which people can someday rally and renew themselves. We can now pull together what is left and hope that it demonstrates the viability of what was given us long ago and that may be sufficient for our life-times.[3]

Deloria's statement may strike some as a desperate appraisal of the remnants of Native North American lifeways. In light of the ecological degradation imposed on the planet and the pervasive economic drive to continually raise gross national product with little or no thought of gross earth product, such an assessment as that by Deloria does not seem unrealistic.[4] What is striking, how-ever, is Deloria's term "viability" to describe North American life-ways. Exploring this capacity for growth and for change within Native North American worldviews is central to this essay. This viability, I would suggest, is not simply a value of the past oral narrative traditions but endures in many contemporary ritual set-tings of indigenous peoples. This essay seeks to explore selected aspects of the viability of Native American lifeways beginning with a traditional Crow-Apsaaloke invocation which provides insight into the basic environmental awareness, or "ecoconsciousness," of Native North American peoples.[5] I will then discuss selected ritu-als, the idea of spirit-presences and environmental ethics as pro-viding viability in the functional cosmologies of Native North American cultures.

Ecoconsciousness in Native American Worldviews

"The sky is my Father and these mountains are my Mother." When I first heard this prayer at a Crow-Apsaaloke *Ashkisshe* cere-mony[6] in 1983 I sensed the endurance of traditional relations between these horse-loving Native peoples and their northern plains homeland. The expansive clear skies of Montana, the state in which the Apsaaloke have their reservation, contrast sharply with the wooded horizon cut by the surrounding Wolf, Big Horn, and Pryor Mountains. In the Apsaaloke worldview the relation-ship established between the human and the presence of the sa-cred is especially focused on the mountains as the sources of fecundity and material blessings.

This relationship between the Crow-Apsaaloke and the moun-tains is especially evident in the "New Year" dance on the last day of Crow Fair, the major reunion, pow wow, and rodeo event in late summer. This dance with its four stops (symbolizing autumn, winter, spring, and summer) is led by a pipe carrier across whose path no human or animal should pass. This honored leader prays that everyone will be healthy for the year and will return again to Crow Fair. People dance behind him in traditional garb; others participate by gathering alongside the dance route. During this dance all the participants gesture in unison toward the mountains as a sign of loving respect for their Mother, the mountains. This is a striking transmission of an ancient worldview value and ritual evocation of human-earth relations in the modern setting of the Crow Fair.

The sacred relationship of the Crow with their mountainous homeland is not articulated in a vague manner but in robust statements of life lived in proximity to the natural world. The nineteenth-century Apsaaloke chief, Arapooish, expressed his understanding of the sacred movement on the land in this manner:

> The Crow country is exactly in the right place. It has snowy mountains and sunny plains; all kinds of climates and good things for every season. When the summer heat scorches the prairies, you can draw up under the mountains, where the air is sweet and cool, and grass fresh and the bright streams come tumbling out of the snowbanks. There you can hunt the elk, the deer and the antelope, when their skins are fit for dressing; there you will find plenty of white bears and mountain sheep. In the autumn, when your horses are fat and strong from the mountain pastures, you can go down on the plains and hunt the buffalo, or trap beaver in the streams. And when winter comes on, you can take shelter in the woody bottoms along the rivers; there you will find buffalo meat for yourselves, and cottonwood bark for your horses; or you may winter in the Wind River Valley where there is salt weed in abundance. The Crow country is exactly in the right place. Everything good is to be found there. There is no country like the Crow country.[7]

An aesthetic charge in this statement flows from its powerful ethic which understands the appropriate use of the resources in the land.[8]

Arapooish's lyrical statement carries a moral subtext that one should move in this country in a manner so as to cultivate all the goodness which can come from this land. What is apparent, however, is that Arapooish is speaking to his own people and in his own time. The sacred goodness of Crow country in that day was in its ability to provide a lifeway for the people. We would not expect this worldview value to be expressed by contemporary traditional Apsaaloke people in the same words and in the same manner as when Arapooish spoke so eloquently. This is the case because much of the Apsaaloke homeland described in this statement has been taken from the Crow through treaties, land purchases, or land thefts. Yet, the prayer to the Sky-Father and Mountain-Mother suggests that the Apsaaloke still speak about their land as sacred despite years of pressure to abandon traditional values and to give up their land.

The Apsaaloke worldview celebrated and fostered in the brief opening prayer, as well as in Arapooish's statement, does not express a scientific-conservationist attitude which seeks to manage nature so as to preserve it for future use. Still, it seems as if the traditional Apsaaloke worldview does not preclude the possibility for any individual member of this nation from developing a scientific concern for preserving species and ecosystems.[9] As the geologist Lauret E. Savoy, of African-American and Native American heritage, has written:

> Earth science is an open inquiry into the workings of nature. As scientists, we must also remember that geologic and geographic setting of regions—the physical landscape—has influenced or controlled patterns of human exploration and settlement, as well as lifeways of peoples and their cultural landscapes. Knowledge of the earth—the "geologic experience"—can have varied expressions beyond systematic scientific analysis. Ideally, a holistic perception of the earth might include an understanding of the general concepts of how the natural world operates, a familiarity with the nature and methods of scientific inquiry, and a realization of the importance of earth science in everyday life and the interdisciplinary relationships between the earth and human existence.[10]

The traditional Apsaaloke worldview orients a person toward an affective, symbolic, mythopoeic, private, and ethical relationship between the human and the sacred realms of the earth and the sky. Such a relationship is uniquely understood and "performed" from different Native American worldviews. For example, the senior anthropologist and Native American from San Juan pueblo, Alfonso Ortiz, has observed:

> Among the many people who subscribe to the belief that four mountains define tribal territory are the Navajos, all of the Pueblos, the Pima, and the Yuman tribes of the Gila River. . . . But mountains are more, much more, than boundary markers defining the tribal

boundaries within which a people lives and carries on most of its meaningful, purposeful activities. The Pueblo people, for instance, believe that the four sacred mountains are the pillars which hold up the sky and which divide the world into quarters. As such they are imbued with a high aura of mystery and sanctity. And this sacred meaning transcends all other meanings and functions. The Apaches, the most recent mountain dwellers among the southwestern Indians, believe that mountains are alive and the homes of supernaturals called "mountain people." They further believe that mountains are protectors from illness as well as external enemies, that they are the source of the power of shamans as well as teachers of songs and other sacred knowledge to ordinary humans, and that, finally, mountains are defenders of tribal territory.[11]

This sensitivity to the vital character of all life as well as an openness to developing techniques of interaction with the natural world is a crucial perspective for an appreciative understanding of both the ecological dimensions and the enduring viability of Native North American lifeways.

Rituals, Spirits, and Ethics in Functional Cosmologies of Native North America

The intimacy of relations between the various indigenous peoples of North America and their bioregions is no longer a vague or moot point.[12] Extensive ethnographic documentation details the symbolic and ritual regard for local bioregions.[13] For example, the late eighteenth- and early nineteenth-century explorations of David Thompson described his encounter with the Salish-speaking, Kettle Falls people along the Columbia River. At that encounter the Kettle Falls villagers insisted that no human beings use the river for washing or toiletry because the salmon people, the symbol of life for many village groups in this region, were spawning.[14] This type of ecoconsciousness demonstrates a relationship that is at once practical and spiritual. The practical relationship evidenced in subsistence hunting, fishing, or gathering was accompanied by a spiritual relationship. This spiritual relationship was typically associated with the "adoption" of the Native person by the spiritual power believed to reside within the animal, fish, or plant. The knowledge which was imparted to the individual hunter, fisherperson, or gatherer was private and secret because it was individually transmitted, often during vision quests or unsolicited visionary experiences or dreams. However, this private and spiritual power was publicly displayed at appropriate moments during the ritual calendar and publicly utilized during subsistence activities coordinated with the seasonal calendar.

Countless examples can be adduced of the Native knowledge of herbs, plants, trees, and fungi which were not simply understood as material techniques but as interactions with living, spiritually empowered beings.[15] In addition to this medical pharmacopoeia and cultivation of plant growth, indigenous peoples of the Americas developed extensive awareness of astronomical activity and animal behavior upon which their livelihood depended. These sacred relationships continue to be experiential and private among indigenous peoples, reflecting both an adaptation to diverse bioregions and an awareness of spiritual dependence upon these local resources as guides into a sacred bond.

In the primal time of myths, one seminal feature of this bond with the natural world is an understanding that animals and plants had sacrificed their bodies so that humans might live. Thus, the ecoconsciousness of many Native North American peoples is directly connected to indigenous rituals in which individuals sacrifice themselves. The somatic deprivation during vision fasting of Plains peoples, as well as the piercing during the Sun Dances, are spiritual acts which are directly tied to specific functional cosmologies. By fasting or by cutting or by marking parts of their bodies, individuals seek vision experiences as a testimony of this reciprocal bond with spiritual powers in the world.

Prominent world renewal ceremonies among different Native American peoples demonstrate profound reflection on the existential challenge of life taking in subsistence practices. Thus, the Winter Dance among the Salish peoples of the Columbia River plateau country occurs when traditional vision singers come together in midwinter to sing the songs which the animals and plants gave to humans in primal, mythic times. These singers speak of a "spirit sickness" which they suffer during the days and weeks before they sing. This spirit sickness is a unique cultural disposition linked to their worldview. Spirit sickness cannot simply be reduced to a psychological behavior. Rather, it is also a contemplative path which, if followed for a lifetime, leads to a deepened understanding of both human transitions in aging and the profound link between the human and nonhuman worlds. The meditative path, which can only be suggested in the phrase "spirit sickness," results from the traditional worldview understanding of the primal sacrifice of the animals and plants. Often associated with specific geographical places or locations on mountains, these mythic beings continue to give of their bodies as well as of their songs, according to this Salish worldview, so that humans might live. During the Winter Dance, the singers and participants not only eat the sacred foods—salmon, deer, bitterroot, and camus—which sustain life, they also sing the songs which provide life direction and reflection on the shared sacrifice and interdependence of all life.[16]

The *Massaum* ceremony of the Tsistsistas-Cheyenne peoples was an event which drew together ancient shamanistic capacities to call animals into a hunter's entrapment with the Tsistsistas memory of seeking ritual validation for their entrance onto the Missouri River drift plains. This ancient ceremony was last performed in the early twentieth century according to ethnographic accounts.[17] The *Massaum* was a complex of rituals which involved masked animal dancers, specially prepared lodges and sacred earth altars which made present cosmic powers, and mythic narratives of the achievements of the culture hero, Motseyoef. This ceremonial was a complex, multilayered symbolic process over several days in which the Tsistsistas celebrated their mythic remembrance of themselves as a people. Most importantly, the *Massaum* contained ritual actions which the Tsistsistas people undertook to legitimate their entry onto the Plains. They evoked the relationship between their culture hero and the animals of the Plains to sanction their presence in this country. Thus, the *Massaum* is a complicated illustration of the "viability" of which Deloria indicated in his remarks above. This richness of experience, belief, performance, ethics, and social cohesion is constellated in the *Massaum* and suggested in the term *lifeway*.

A central symbolic axis of the *Massaum* ceremony as an "earth-giving" ceremony was grounded in the "earth-mound altars." These altars were prepared from layers dug out of the surface soil where humans walk and from the "deep earth" where the spiritual powers which sustain human life are believed to reside. The mound altars symbolically make present the interior forces of the mythic cave in which Motseyoef encountered the powers of the earth. These powers *(vonoom)* in the earth, sky, and deep regions of both earth and sky were evoked during *Massaum* as an act of renewing the primordial tie of the Tsistsistas people with the primal order (the same term, *vonoom*, is used for this original order) in the cosmos. In the manifold ritual acts of *Massaum*, then, was embedded an ecoconsciousness in which these people validated their habitation on the northern grasslands of the continent by appealing to the animals, plants, and geographical places of their northern plains homeland.[18] One could dwell on the poignant loss of such a profound ritual evocation of human-earth relations or, conversely, hope and encourage the Tsistsistas in their contemporary effort to transmit the values embedded in that ritual.[19]

Perhaps one of the most problematic worldview values of indigenous peoples according to those coming from the Western monotheistic traditions is the conception of a plurality of spiritual powers in the landscape. From the standpoint of the indigenous worldview of sacrifice and dependence upon local bioregions, the plurality of spirits in nature manifests the bountiful gifts of the natural world. The ritual giveaway, so prominent across Native North America, is a clear example of a ritual act which transmits ethical values based on a functional cosmology of respect for the earth's bounty.

Many Native North American peoples have clearly developed concepts of an overarching creative presence such as *Wakan Tanka* and *Kitche Manitou*. These conceptions of a unified relationship throughout the plurality of spirit-presences in the landscape demonstrate a significant dynamic of Native North American religious thought. Not simply philosophical reflection on the One and the Many, these concepts of a "great mysterious presence" as creator underscore the synthetic whole to which all life is believed to relate. The plurality of spirits, for example, appears to be the diverse manifestations of a singularly mysterious and cosmic power.[20] Often labeled as belief in a "high god," this type of god-interpretation draws on Atlantic-Mediterranean biblical taxonomies to organize and explain indigenous lifeways. These interpretive postures have not only found their way into ethnographic texts describing traditional thought; they have also penetrated into the religious life of Native North American peoples today, for example, the Christianized forms of the Peyote Way. The actual meaning of such numinous presences as *Wakan Tanka* or *Kitche Manitou* and their relationship to the layering of "power" as *wakan* or *manitou*—that is, "spirits" in the landscape—remains the interpretive, experiential, and private realm of contemporary traditional practitioners.

In addition to these functional, cognitive, and contemplative systems in which the ecological dimensions of Native North American religions are organized, there are highly developed mythopoeic traditions which are anchored in traditional places sacred to the personal and social memory of specific indigenous groups. The complex interweaving of kinship, myth, ethics, and bioregion is poignantly displayed in the Apache peoples' belief that a mountain can "stalk" and "shoot" a person with stories. The anthropologist Keith Basso has described this aspect of the Apache lifeway, saying:

> Mountains and arroyos step in symbolically for grandmothers and uncles. Just as the latter have "stalked" delinquent individuals in the past, so too particular locations continue to "stalk" them in the present. Such surveillance is essential, Apaches maintain, because "living right requires constant care and attention," and there is always a chance that persons who have "replaced themselves" once . . . will relax their guard against "badness" and slip back into undesirable forms of social conduct. Consequently, Apaches explain, individuals need to be continuously reminded of why they were "shot" in the first place and how they reacted to it at the time. Geographical sites, together with the crisp mental "pictures" of them presented by their names, serve admirably in this capacity, inviting people to recall their earlier failures and encouraging them to resolve, once again, to avoid them in the future. Grandmothers and uncles must perish but the landscape endures, and for this the Apache are deeply grateful. "The land," Nick Thompson [an Apache elder] observes, "looks after us. The land keeps badness away."[21]

The extensive hunting and agricultural prohibitions, often dismissed as superstitions in overly rational interpretations, constitute a practical form of environmental ethics. These indigenous environmental ethics flow from specific worldviews and respond to specific peoples' dispositions to act in relationship to a living sacred world. The oral narratives, or mythologies, which describe these relationships also evoke the spiritual relationship itself. Thus, a Koyukon elder reflected on the larger meaning of the names of animals within his northern boreal forest home:

> The animal and its spirit are one in the same thing. When you name the animal you're also naming its spirit. That's why some animal names are *hutlaane* [prohibited, taboo]—like the ones women shouldn't say—because calling the animal's name is like calling its spirit. Just like we don't say a person's name after they die . . . it would be calling their spirit and could be dangerous for whoever did it.[22]

Richard Nelson's investigations of the Koyukon attitude toward their resources brought him into an understanding of these *hutlaane,* or prohibitions, as an elaborate ethical system for managing and protecting the bioregion from human exploitation. He comments that the Koyukon "have developed an ethic of conservation, manifested in concepts of territory and range, attitudes towards competitors for subsistence resources, methods of avoiding waste, and implementation of sustained yield practices."[23] While Nelson uses the term "conservation," obviously this ethic flows from, and affirms, worldview values quite different than scientific conservation.

Other Native North Americans also use evocative kinship terms to signal deeper relationships with their environment. This is stated eloquently by Chief Oren Lyons, Faithkeeper of the Onondaga Nation, which is within the Houdenosaunee Confederacy of Northeastern North America:

> In the perception of my people, the *Houdenosaunee*, whom you call Iroquois, all life is equal, and that includes the birds, animals, things that grow, things that swim. It is the Creator who presents the reality. As you read this by yourself in your sovereignty and in your being, you are a manifestation of the creation. You are sovereign by the fact that you exist. This relationship demands respect for the equality of all life.
>
> During the 500 years we have interacted with Euro-Americans over 400 treaties have been exchanged between Indian Nations and the United States. In fact, the Iroquois Confederacy, known as the Great League of Peace, was influential in the drafting of the documents which founded the United States. Absent from the political thinking in the United States, however, has been an understanding of the equality of all life and a perspective for nurturing future life. This respect for future life, in my people's understanding, demands that we look ahead. In all decision-making we consider: will this decision be to the benefit and welfare of the seventh generation? Now, it is time for the indigenous peoples to speak about that which we have observed— exploitation of not only the people but also of the earth's resources without any regard for the seventh generation. Caring for the earth, then, calls for sovereign responsibility not simply to yourselves, but to your people, your earth, your seventh generation.[24]

Indigenous traditions raise environmental issues in a cosmological context which includes use of resources but does not situate the discussion primarily in either transcendental or strictly utilitarian terms. By "transcendental" I refer to the position of some religions in which matter and the material world of resources are seen as base or lower than an ultimate, salvific experience beyond this phenomenal realm. The religious utilitarian position understands material resources as simply given by the divine for human exploitation with little reflection on the relation of cosmology to resource exploitation.

As is evident in the preceding remarks, the religious values within the lifeways of Native American peoples, which have extensive similarities with other global indigenous peoples, serve to enrich the ecological dialogue in assessing the importance of functional cosmologies, environmental ethics, and appropriate rituals for renewing human-earth relations. The ecological dimensions of Native North American lifeways are not held back in some evolutionary backwardness or in a stereotyped ahistorical paradise. Nor can the spiritual insights of the First Peoples of the North American continent be descried as lacking profundity. Moreover, the private and experiential nature of indigenous traditions does not lead to a muddled spiritual life but rather one in which individual values are embedded in a lifeway community which extends into the natural world. In an investigation of these issues, striking insights emerge which can complement, correct, and enrich the very traditions which often dominate these indigenous traditions.

Openness to authentic Native North American voices brings to the ecological dialogue such insights as those articulated by Richard West, Jr., Director of the Smithsonian's National Museum of the American Indian, who is also a member of the Cheyenne-Arapaho tribe of Oklahoma. He was interviewed by the Modoc Indian writer, Michael Dorris, on the topic of Native American art:

> ". . . if you avoid being obsessed with the minutiae of the material and look at the 'surround,' whether it's philosophical, ideological or intellectual, then you get beyond the specific and into the realm of ideas. You begin to touch on legitimate points of communality. It seems to me that the guiding set of esthetics in [American] Indian art is inextricably tied to a shared concept of nature and a belief that life exists in things others might see as inanimate."
>
> This perspective, Mr. West believes, results in an attitude toward the environment peculiar to tribal peoples everywhere and its celebration may well address a spiritual void within the industrialized world.[25]

The ecological dimensions of Native North American religions cause us to reconsider our basic relatedness to the world that sustains us in terms of each particular cosmology. These Native American religious values cover a range of human experiential and social activities embedded in oral narratives, ritual practices, and cosmological understandings. It seems apparent that the introduction of indigenous thought traditions into the ecological dialogue would be of vital importance in relation to the current degraded state of regional environments around the globe.

Notes

1. On 5 November 1992 Chief Oren Lyons, Faithkeeper of the Onondaga Nation, which is within the Houdenosaunee Confederacy of Northeastern North America, gave this figure for North America.

2. A noticeable difference in the way that cosmologies function is evident in place-names throughout America. The retention of Indian place-names, which carried cosmological charge, into the "meets and bounds" worldview of contemporary American property rights serves to illustrate the ongoing appropriation of Native American worldview values. See John Rydjord, *Indian Place-Names* (Norman: University of Oklahoma Press, 1968), passim.

3. Vine Deloria, Jr., "Is Religion Possible? An Evaluation of Present Efforts to Revive Traditional Tribal Religions," *Wicazo Sa Review* 8, no. 1 (Spring 1992): 35–39.

4. See Thomas Berry, *The Dream of the Earth* (San Francisco: Sierra Club Books, 1988).

5. The term *ecoconsciousness* in this context is developed in *Handbook of American Indian Religious Freedom*, ed. Christopher Vecsey (New York: Crossroads Press, 1991).

6. The Crow people, who call themselves *Apsaaloke* or *Absaroke* in their own language, which is still vital and functional, have a reservation in south central Montana in the United States. The author has been visiting Apsaaloke families and friends since the early 1980s. Integral to understanding Apsaaloke religious attitudes toward their homeland is an appreciation of the varied Apsaaloke sacred calendars. For example, the *Ashkisshe*, known in English as the Sun Dance, is connected with the seasonal renewal of community life in the spring. The *Ashkisshe* is a world renewal ceremony in which dancers enter a specially prepared lodge and undergo a fast from food and water in order to pray for their community of all living beings—human, animal, plant, mineral, and geographical. Importantly, the Christian calendar is also known and widely practiced as are ritual calendars associated with the unique Apsaaloke ceremonies of The Tobacco Society. For background on the Crow-Apsaaloke see Rodney Frey, *The World of the Crow Indians* (Norman: University of Oklahoma Press, 1987).

7. Quoted in J. Donald Hughes, *American Indian Ecology* (El Paso: Texas Western University Press), 142; originally in Stewart L. Udall, *The Quiet Crisis* (New York: Holt, Rinehart & Winston, 1963), 17–18.

8. It is interesting in light of this ecological dialogue to reconsider older sources on mountains such as W. Y. Evans-Wentz, *Cuchama and Sacred Mountains*, ed. Frank Waters and Charles Adams (Chicago: Swallow Press, 1981).

9. Native American conservationists are discussed in William Stolzenburg, "Sacred Peaks, Common Grounds: American Indians and Conservationists Meet at a Cultural Crossroads," *Nature Conservancy* (September/October 1992): 16–23; for a differing perspective see an interview with Nicanor Gonzalez, "We Are Not Conservationists," *Cultural Survival* 16, no. 3 (Fall 1992): 43–45.

10. Lauret E. Savoy, "Encounters with the Land," *GSA Today: A Publication of the Geological Society of America* 2, no. 10 (October 1992): 215–18.

11. Alfonso Ortiz, "Look to the Mountaintop," *Essays on Reflection*, ed. Ward E. Graham (Boston: Houghton-Mifflin, 1973), 91–92.

12. In addition to the work by J. Donald Hughes cited in n.7, see *American Indian Environments: Ecological Issues in Native American History,* ed. Christopher Vecsey and Robert Venables (Syracuse, N.Y.: Syracuse University Press, 1980); O. Douglas Schwarz, "Indian Rights and Environmental Ethics: Changing Perspectives and a Modest Proposal," *Environmental Ethics* 9 (Winter 1987): 291–302; Annie L. Booth and Harvey M. Jacobs, "Ties That Bind: Native American Beliefs as a Foundation for Environmental Consciousness," *Environmental Ethics* 12 (Spring 1990): 27–43; and Saroj Chawla, "Linguistic and Philosophical Roots of Our Environmental Crisis," *Environmental Ethics* 13 (Fall 1991): 253–62.

13. One example of this growing body of documentation is James V. Spickard, "Environmental Variation and the Plausibility of Religion: A California Indian Example," *Journal for the Scientific Study of Religion* 26 (1987): 327–29. However, lest this discussion be cast simply in a consideration of early ethnographic documentation, consider the contemporary political-legal fight of the Taos people for custody and ritual access to their sacred Blue Lake in R. C. Gordon-McCutchan, *The Taos Indians and the Battle for Blue Lake* (Santa Fe, N.M.: Red Crane Books, 1991).

14. For David Thompson reference see Richard Glover, ed., *David Thompson's Narrative: 1784–1812* (Toronto: Champlain Society, 1962), xcviii and 335–58.

15. For example see Nancy Turner, Randy Bouchard, and Dorothy Kennedy, *Ethnobotany of the Okanagan-Colville Indians of British Columbia and Washington,* No. 21, Occasional Paper Series, British Columbia Provincial Museum, Victoria, 1980.

16. See John A. Grim, "Renewing the Earth: Religion and Ecology in the Winter Dance of the Kettle Falls People," *Bulletin Pontificium Consilium pro Dialogo Inter Religiones* 27, no. 1 (1992): 65–89.

17. See George Bird Grinnell, *The Cheyenne Indians,* 2 vols. (New Haven: Yale University Press, 1923).

18. Karl H. Schlesier, *The Wolves of Heaven: Cheyenne Shamanism, Ceremonies, and Prehistoric Origins* (Norman: University of Oklahoma Press, 1987).

19. Outsider-ethnographers have mistaken the demise of such rituals as the Ojibway Midewiwin when in fact they continue to be performed; see Christopher Vecsey, *Traditional Ojibwa Religion and Its Historical Changes* (Philadelphia: American Philosophical Society, 1983).

20. See Ake Hultkrantz, *The Religions of the American Indians* (Berkeley: University of California Press, 1967), 15–26.

21. Keith H. Basso, "'Stalking with Stories': Names, Places, and Moral Narratives among the Western Apache," in *Text, Play, and Story: The Construction and Reconstruction of Self and Society,* ed. E. Brunner (Washington, D.C.: American Ethnological Society, 1984), 112–13.

22. Richard K. Nelson, *Make Prayers to the Raven: A Koyukon View of the Northern Forest* (Chicago: University of Chicago Press, 1983), 22.

23. Ibid., 216.

24. From a public address by Chief Oren Lyons at the Weis Performing Arts Center, Bucknell University, Lewisburg, Pa., 5 November 1992.

25. Michael Dorris, interview with Richard West, *The New York Times,* 13 September 1992, 53.

Judaism and the Ecological Crisis

Eric Katz

New Jersey Institute of Technology

W HAT does Judaism say about nature and the environmental crisis? Any discussion of the Jewish view of the natural
world, the ecological principles underlying natural processes, and
the obligations relevant to human activity in relation to nature
must begin with the concrete and specific commandments binding
upon all practicing Jews. A so-called worldview of Judaism would
be a mere abstraction from the specific rules and principles of
Jewish life, for in Judaism, perhaps more than any other religion,
philosophical meaning arises out of the procedure of concrete
daily activity. As Robert Gordis writes: "The true genius of Judaism has always lain in specifics." Thus, Gordis continues, an understanding of Jewish teachings on the environmental crisis is
"not to be sought in high-sounding phrases which obligate [Jews]
to nothing concrete; rather [it] will be found in specific areas of
Jewish law and practice."[1]

This reluctance to focus on abstract philosophical principles or
a generalized worldview as a replacement for concrete obligations
regarding the natural environment is a recurrent theme in the
expositions of contemporary commentators on the Jewish tradition. After a discussion of the Hebrew concepts of nature in the
Bible, Jeanne Kay concludes, in part, that "the Bible views observance of its commandments, rather than specific attitudes toward
nature or techniques of resource protection, as the prerequisite
of a sound environment."[2] In an essay that predates the current
environmental crisis by more than a generation, E. L. Allen claims
that in the Jewish tradition nature is neither an abstraction nor
an ideal, but rather one of the realms in which humans interact
with God. "Nature is envisaged as one of the spheres in which
God meets man personally and in which he is called upon to
exercise responsibility."[3] Thus, "for the man of the Bible nature
is never seen in abstraction either from God or from the tasks
which He has assigned to man in the world."[4] Within Judaism,

55

then, the human view of nature and the environment is grounded
in the specific obligations and activities of Jewish life, the tasks
and commandments that God presented to the Jewish people.

Subdue the Earth: Dominion and Stewardship

Given this turn away from the abstract, an examination of the
Jewish perspective on nature and the environmental crisis must
begin with specific texts and commands, and none is more important than Genesis 1:28 in which God commands humanity to
subdue the earth:

> And God blessed them [Adam and Eve]; and God said unto them:
> "Be fruitful and multiply, and replenish the earth, and subdue it; and
> have dominion over the fish of the sea, and over the fowl of the air,
> and over every living thing that moves upon the earth."

This notorious passage appears in almost every discussion of the
religious foundations of the environmental crisis. It is used by
Lynn White, Jr., and others, to demonstrate that the Judeo-Christian tradition is fundamentally biased toward the dominion—if
not the actual domination—of the earth by humanity.[5] It suggests
that the earth and all nonhuman living beings in nature belong
to the human race, mere means for the growth ("be fruitful and
multiply") of humanity.

This is not the place for a full discussion of White's controversial
thesis concerning the Christian tradition.[6] But if we are to understand the Jewish perspective on the environmental crisis, we must
examine the meaning of the command to "subdue" the earth and
its relationship to the process of domination. Does this passage
represent God's gift of title to humanity? Does this passage mean
that the earth belongs to the human race?

The Jewish tradition clearly answers in the negative. Norman
Lamm points out that the very next line from Genesis, which is
usually ignored in the discussions of this passage, restricts humans
to a vegetarian diet, hardly the prerogative of one who has dominion, control, and ownership of all the living creatures in nature!
"And God said: 'Behold, I have given you every herb yielding
seed, which is upon the earth, and every tree in which is the fruit
of a tree yielding seed—to you shall it be for food'" (Gen. 1:29).

The Torah thus limits the human right to "subdue" and use nature; this command is not title to unbridled domination.[7]

Indeed, Jewish scholars throughout history have gone to extraordinary lengths to disavow any idea that Genesis 1:28 permits the subjugation of nature by humanity. The Talmud (*Yebemot* 65b) relates the phrase "subdue it" to the first part of the sentence, "be fruitful and multiply," and then through a tortuous piece of logic, connecting the act of "subduing" with warfare—a male activity—claims that the passage really means that the propagation of the human race is an obligation of the male. And the medieval commentators Nachmanides and Obadiah Sforno connect the phrase to the activities of humanity in the use of natural resources, not their destruction or misuse. Nachmanides sees the passage as granting permission to humanity to continue their activities of building, agriculture, and mining. Sforno's explanation is even more restrictive: "*And subdue it*—that you protect yourself with your reason and prevent the animals from entering within your boundaries and you rule over them."[8] These interpretations recognize the power of humanity to use natural resources, and indeed the necessity of them so doing, but they emphasize limitations in the human role. Dominion here does not mean unrestricted domination.

The reason for the restrictions is also clear: in the Jewish tradition, humanity is the steward of the natural world, not its owner. Stewardship is a position that acknowledges the importance of the human role in the care and maintenance of the natural world without permitting an unrestricted license. David Ehrenfeld and Philip Bentley thus consider it a middle position, but one that is definitely on the side of the spectrum that advocates the human use of the natural environment, rather than the opposite extreme of the sacred reverence and noninterference with nature suggested by Eastern religions such as Jainism.[9] To use a comparison widespread in the literature of environmental philosophy, the concept of stewardship in Judaism advocates neither the *domination-destruction* nor the *preservation* of the natural environment but its *conservation* and wise developmental use. Genesis 2:15 lends support to the idea of stewardship, as it declares: "And . . . God . . . put him into the garden of Eden to till it and to keep it." This suggests, as Ehrenfeld and Bentley point out, that the human dominion over nature should not be interpreted in a harsh or exploitative way, and the rabbinic tradition has not done so.[10] The whole idea of stewardship implies care for an entity that is in one's power; it does not imply exploitative use.

The idea of care implicit in stewardship is, however, based on a more fundamental concept: the proper ownership of the entity under care. From a mere analysis of the meaning of concepts, the difference between dominion and stewardship is that the former includes an unrestricted ownership and total power over the subordinate entity, while the latter strictly limits power because it denies ownership. Humanity does not own the natural world. *In Judaism, the world belongs to God.* Judaism is a theocentric religion, at least when it concerns the relationships between humans and nature. God himself, not human life and welfare, is the source of all religious and moral obligation. The divine ownership of nature is most clearly and directly stated in Psalm 24: "The earth is the Lord's and the fulness thereof, the world and those who dwell therein." Humanity cannot have an unrestricted dominion over the natural world because the world belongs to God; humanity is merely the divinely appointed guardian or steward of what belongs to God.

This general theocentric worldview is expressed in many ways throughout Jewish ritual and practice, so much so that Jonathan Helfand can declare that "in both content and spirit the Jewish tradition negates the arrogant proposal that the earth is man's unqualified dominion."[11] God does not forsake the ownership of the world when he instructs Adam and his descendants to master it. As Helfand notes, the existence of the laws concerning the sabbatical—and the Jubilee—year clearly indicate that God is the owner of the earth: "And the land is not to be sold in perpetuity, for all land is Mine, because you are strangers and sojourners before Me" (Lev. 25:23).[12] In Samuel Belkin's words, man possesses but a "temporary tenancy of God's creation."[13] Thus, the prohibition on farming the land in the seventh year, which is detailed in Leviticus 25:3–4, is not to be understood as a primitive attempt at enlightened agricultural methods. Belkin argues that "the sages refuse to assign purely economic, agricultural or social motives to this law," for Rabbi Abahu cites the ownership of God as the primary reason for the existence of the Sabbath and Jubilee years (*Sanhedrin* 39a).[14] Belkin himself is even more emphatic about the theocentrism of Judaism: "the entire structure of Judaism rests" on the principle "that creation belongs to the Creator." Without such a principle, humans would own the world and the entities within it; they would then be able to use those things without regard to any laws or principles other than their own will. But this is not the case: the moral code of the Torah, the ritual commands, and the laws of Judaism all strongly imply that the world belongs to God, and he has "instructed man concerning what he is permitted to do or prohibited from doing with His creation . . . [God] alone dictates the terms of man's tenancy in this world."[15]

One commonplace example of the way ritual action reinforces the notion of God's ownership is the commandment concerning the blessings over food. Helfand cites the *Tosefta: Berakhot* 4:1:

> "Man may not taste anything until he has recited a blessing, as it is written: 'The earth is the Lord's and the fulness thereof.' Anyone who derives benefit from this world without a [prior] blessing is guilty of misappropriating sacred property."[16]

The fact that God owns the world requires us to ask permission before we ingest any item of food. All the objects of the material world are as sacred as the entities of heaven, for they are all the creation of God, and belong to him.[17]

Perhaps the significance of the theocentric ownership of the world by God in Judaism is best summarized by the rituals concerning not the sabbatical year, but the ordinary *weekly* Sabbath. Ehrenfeld and Bentley articulate the meaning of the Sabbath for contemporary environmentalists: "For Jews, it is the Sabbath and the idea of the Sabbath that introduces the necessary restraint into stewardship."[18] For these authors, the Sabbath acquires this meaning because of three elements of the observance of Sabbath: "we create nothing, we destroy nothing, and we enjoy the bounty of the Earth." The fact that nothing is created serves to remind us that we are not as supreme as God; the fact that nothing is destroyed emphasizes that the world does not belong to us, but to God; and our enjoyment of the earth's bounty reminds us that God is the source of nature's goodness.[19] Thus the concept of the Sabbath itself—the absence of work and the appreciation of God—imposes a strict limit on human activity and achievement. Humanity in no way possesses dominion over the nonhuman world since it does not even possess dominion over its own activities.

Observance of the Sabbath thus returns us to the notion of stewardship, for without dominion, humanity is merely the steward of God's creation. But stewardship strongly implies a notion of responsibility, for the steward is responsible for the condition of the entities in his care. To illustrate this point, Ehrenfeld and Bentley recount a story told by the eleventh-century Spanish rabbi, Jonah ibn Janah. A man walks into a house in the midst of a deserted city; he finds a table with food and drink and begins to eat, thinking to himself, "I deserve all this, it is mine, I will act as I please." Little does he know that the owners are watching him, and that he will have to pay for all that he consumes. Thus man, as merely the appointed steward of God's creation, is responsible to God for the use of his property, the natural world.[20]

Environmentalism in Practice: Rituals and Commandments

An abstract notion of responsibility for the guardianship of the natural world is not, however, an adequate account of Judaism's perspective on the environmental crisis. For this notion of responsibility to be part of the practice of religious belief, it must be distilled into a series of specific commandments regarding human actions affecting the natural world. An examination of Jewish law and ritual does reveal these specific commandments, involving many different aspects of everyday Jewish life.

Several commandments involve the general health and well-being of the human community as it is situated in the natural environment. Deuteronomy 23:13–15, for example, requires the burial of human sewage in wartime, with the command that the soldiers must possess a spade for that very purpose among their other weapons: "and it shalt be when thou sittest down outside, thou shalt dig therewith, and shalt turn back and cover that which cometh from thee."[21] A more general principle is yishuv ha-aretz ("the settling of the land") which mandates both restrictions on the type of animals that can be raised and the type of trees that could be used for burning on the sacrificial altar. Goats and sheep were thought to be destructive to the land, and vine and olive trees were too valuable to be used in religious services.[22] Helfand argues that yishuv ha-aretz is also the basis of the mandate to establish a migrash, an open space one thousand cubits wide around all cities in Israel, in which agriculture and building would be prohibited. "The operative principle ... calls upon the Jew in his homeland to balance the economic, environmental, and even religious needs of society carefully to assure the proper development and settling of the land."[23]

The existence of the migrash is, indeed, only one aspect of the laws regulating life in early Jewish cities, what amounts to a fully realized notion of town planning. Aryeh Carmell discusses many of these restrictions in an essay detailing the rabbinic concern for the quality of the environment in Jewish life.[24] Rambam in the *Hilchot Shechenim* ("Laws of Neighborly Relations"), explains that there are four classes of nuisance in which injury is always presumed: smoke, dust, noxious smells, and vibration. There is also a right to quietness.[25] This leads to rabbinic regulations—a kind of ancient zoning ordinance—regarding the specific placement of certain "industries" within the town: threshing floors, cemeteries, tanneries, and slaughterhouses.[26] The basis of these rabbinic regulations was a limitation of individual property rights for the sake of the entire community.[27] It seems clear that these limitations of individual property rights can also be traced to the notion that all property belongs ultimately to God, and thus that the use of the property by human individuals must be regulated by the laws of the Torah and the rabbinical interpretations of these laws.

Another category of Jewish law concerns the human relation to the divine plan. Nature is conceived, in Judaism, as the result of a divine plan or intelligence, which is not to be altered by human activity. Thus, in Leviticus 19:19 we find a prohibition against the hybridization of plants and animals, and even a restriction on wearing two types of cloth: "you shall not let your cattle mate with a different kind, you shall not sow your field with two kinds of seed, you shall not wear a garment of wool and linen." Helfand explains that this passage falls in the midst of a discussion of the proper and improper forms of human relationships, thereby reinforcing the idea that there is a fixed divine plan for both the social and the natural order of the universe.[28] The intrinsic significance of the divine plan is further revealed by Jewish traditions that aim, in modern terminology, to protect endangered species. Thus Helfand cites the commentator Nachmanides on the meaning of two biblical commands—not to slaughter a cow and her calf on the same day (Lev. 22:28) and not to take a mother bird with her young (Deut. 22:6): "Scripture will not permit a destructive act that will cause the extinction of a species."[29]

A concern for animals is further emphasized in Jewish thought by the fundamental principle of *tza'ar ba'alei chayim* ("the pain of living creatures"). Although it is not strictly a principle concerning the ethical treatment of the environment, it is the basis for the compassionate treatment of animals throughout Jewish life. Gordis considers it one of the two basic principles constituting the Jewish attitude to the nonhuman natural world—the second principle, *bal tashchit*, will be discussed below.[30] *Tza'ar ba'alei chayim* requires a concern for the well-being of all living beings—if not a full-scale sacred reverence for all life, at least an attitude of universal compassion.[31] The laws of kosher slaughtering, as well as the law forbidding the yoking together of animals of unequal strength (Deut. 22:10), are based on this compassion for animal suffering.

As Gordis emphasizes, one of the most unlikely textual affirmations for *tza'ar ba'alei chayim* is the conclusion of the book of Jonah, in which Jonah complains to God about the destruction of a gourd, a plant that had been shielding Jonah from the sun as he awaited God's decision about the destruction of the city of Nineveh. Jonah is angry for two reasons: God has spared the city, thereby making Jonah's prophecy appear foolish or pointless; and God has caused the gourd that shaded him to wither and die. God's reply is this:

> "You pity the gourd, for which you did not labor, nor did you make it grow, which came into being in a night, and perished in a night. And should I not pity Nineveh, that great city, in which there are more than a hundred and twenty thousand persons who do not know their right hand from their left, and also much cattle?" (Jon. 4:9–11)

God's rebuke compels a consideration of three different kinds of entities: the human inhabitants of Nineveh, the nonhuman domesticated animals that live in Nineveh, and the wild gourd—the plant life—outside the city. Clearly God does not consider the potential loss of the cattle to be a minor point; the loss of the cattle with the human population is an event to pity, an event requiring divine compassion. But the passage also suggests that pity for the gourd—wild, undomesticated plant life—is not an absurdity. Jonah's mistake is not that he felt compassion for the gourd, but that his level of concern was too great. It is wrong to value the wild gourd more than God values the inhabitants of Nineveh. Compassion for all living beings is a moral obligation in Judaism, but the context will determine the appropriate level of response.

Bal Tashchit: Do Not Destroy

Although the preceding section listed several principles and commandments that prescribe specific actions regarding the nonhuman environment, the most important and fundamental principle of the Jewish response to nature is *bal tashchit*—"do not destroy"—which is first outlined in Deuteronomy 20:19–20:

When you besiege a city for a long time . . . you shall not destroy its trees by wielding an ax against them. You may eat of them, but you may not cut them down. Are the trees in the field men that they should be besieged by you? Only the trees which you know are not trees for food you may destroy and cut down, that you may build siege-works against the city.

In the context of warfare, specific moral rules apply. As Gordis notes, "this injunction ran counter to accepted procedures in ancient war," particularly the actions of the ruthless Assyrians.[32] But more importantly, the principle of *bal tashchit* forbids the wanton destruction of an enemy's resources, a so-called "scorched earth" policy of warfare. Lamm comments that "what the Torah proscribed is not the use of the trees to win a battle, which may often be a matter of life and death, but the wanton destruction of embattled areas, so as to render them useless to the enemy should he win."[33]

The principle here is the prohibition on wanton destruction or vandalism, the destruction of trees for no (or little) redeeming purpose. Lamm also notes that Jewish law extends the law to situations in peacetime as well as war; the Bible merely used an example of a situation in wartime to emphasize the seriousness of the restriction, for the commandment "do not destroy" is so powerful that it cannot even be overridden for the sake of victory in war.[34] Thus both Lamm and Gordis claim that *bal tashchit* is the establishment of a general principle in the expression of a concrete situation.[35]

There is much evidence from rabbinical texts to support the idea that *bal tashchit* is a general and fundamental principle regarding human actions within the nonhuman and natural environment. The idea of "wielding an ax" is extended to any means of destruction, even the diverting of a water supply.[36] Moreover, the principle is extended to any natural entity or to any human artifact. In the *Sefer Hahinuch* ("Shoftim" Commandment 529) is written this comment on *bal tashchit:* "In addition [to the cutting down of trees] we include the negative commandment that we should not destroy anything, such as burning or tearing clothes, or breaking a utensil—without purpose." Lamm also cites Maimonides, who includes the stopping of fountains, the wasting of food, or "wrecking that which is built" as violations of *bal tashchit.*[37] Thus Gordis concludes: "The principle of *bal tashchit* entered deep into Jewish consciousness, so that the aversion to vandalism became an almost psychological reflex and wanton destruction was viewed with loathing and horror by Jews for centuries."[38]

The precise meaning of *bal tashchit* and its application in the affairs of humans interacting with and using natural objects raises, however, interesting issues. First is the relationship of *bal tashchit* to economic considerations. The original passage in Deuteronomy appears to make a distinction between food-producing (fruit-bearing) trees and trees that do not produce fruit. Although wanton destruction is prohibited regarding all trees, fruit-bearing trees should be protected even from appropriate military uses. It is permitted to destroy trees that do not produce fruit for good reasons. Lamm explains that this special concern for food-producing trees may be tied to commercial considerations, either "an economy of scarcity" or the existence of property rights. And there is rabbinical evidence for the importance of economic values: a fruit-bearing tree may be destroyed if the value of its crop is less than the value of the lumber the tree would produce; moreover, the tree may be destroyed if the land is needed for the construction of a house. These exceptions to *bal tashchit* are not permitted for purely aesthetic reasons, such as landscaping.[39] Eric Freudenstein echoes this conclusion (which he derives from *Baba Kama* 91b): "the standards of bal tashchit are relative rather than absolute. The law is interpreted in the Talmud as limited to purposeless destruction and does not prohibit destruction for the sake of economic gain."[40] But Freudenstein supplements this conclusion with the point that what constitutes an appropriate economic value differs from generation to generation, and thus the correct use of *bal tashchit* at any time must be left to the authorities to decide. The keeping of goats and sheep was once banned because of the destructive impact on the environment, but it is now permitted.[41] Thus the moral evaluation for the destruction of an object or natural entity will depend on the economic and social context of the act. *Bal tashchit* prohibits *wanton* destruction, but the meaning of "wanton" will change throughout history.

An additional economic issue is the relationship of *bal tashchit* to notions of private property. Both Lamm and Gordis claim that the principle is not tied in any way to our modern notion of private property; one is not permitted to destroy one's own property any more than one is permitted to destroy another's. *Bal tashchit* is concerned with "the waste of an economic value per se," i.e., the social utility of the object being destroyed. Lamm even cites the interpretation of the principle to include the idea that it is permissible to destroy a fruit tree if it is somehow damaging the property of others—thus the basis of the principle would be social concern. *Bal tashchit* is a religious and moral law that requires a consideration of the social implications of actions that harm nonhuman entities; it is not a law of financial and personal property.[42]

But even this focus on social consequences does not reveal the true depth of *bal tashchit*. Questions of private property and social utility reintroduce the issue of the real ownership of the world. It was noted above that the fundamental basis of the idea of stewardship was the theocentric perspective of Judaism: the world belongs to God. When *bal tashchit* is combined with this theocentrism, we arrive at the ultimate argument against the destruction of natural entities: such entities are the property of God. This position easily renders insignificant the economic or utilitarian justifications for *bal tashchit*. The principle is not designed to make life better for humanity; it is not meant to insure a healthy and productive environment for human beings. In the terminology of environmental philosophy, it is not an *anthropocentric* principle at all: its purpose is not to guarantee or promote human interests. The purpose of *bal tashchit* is to maintain respect for God's creation.

Gordis thus ties *bal tashchit* to the laws of the sabbatical year and the Jubilee year—the reaffirmation of God's ownership of the land.[43] But as an explanation of the philosophical worldview that underlies *bal tashchit,* I find sections of the book of Job even more compelling. Near the end of the story, Job is finally able to question God about the reasons for the several misfortunes that have befallen him. God speaks to Job out of the whirlwind, but his answer is not a direct justification of the seemingly incomprehensible divine actions that have radically altered Job's life. Instead, God discusses aspects of the natural world—the wild domain outside of human control—and challenges Job to acknowledge the limits of human wisdom:

> Where wast thou when I laid the foundations of the earth?
> Declare, if thou hast the understanding.
> Who determined the measures thereof, if thou knowest?
> Or who stretched the line upon it?
>
> (Job 38:4–5)

And God continues to paint a picture of a world that exists independent of human concerns and free from human notions of rationality or cause and effect:

Who hath cleft a channel for the waterflood,
Or a way for the lightning of the thunder;
To cause it to rain on a land where no man is,
On the wilderness, wherein there is no man;
To satisfy the desolate and waste ground,
And to cause the bud of the tender herb to spring forth?
(Job 38:25–27)

And more than the useless rain on land where humans do not live, there are the animals, the great beasts "behemoth" and "leviathan," which do not exist for human purposes; they lie outside the sphere of human life (Job 40:15ff).

God's speech to Job out of the whirlwind is a dramatic reaffirmation of the theocentrism of the universe, God's creation. Job, as well as any other human being, errs when he believes that the events of the world must have a rational explanation relevant to human life. The events of the world are ultimately explained only in reference to God. This theocentrism is the driving force of *bal tashchit,* for it gives meaning to the reasons behind a prohibition on wanton destruction. Destruction is not an evil because it harms human life—we humans should not believe that God sends the rain for us—it is an evil because it harms the realm of God and his creation.

The remarkable philosophical conclusion from this perspective of theocentrism is that it serves to resolve a long-standing dispute among secular environmental philosophers: should anthropocentric (i.e., human centered) or nonanthropocentric arguments be used to support environmental practices? Should policies of environmental preservation be pursued because such policies will benefit humanity, or because such policies are *intrinsically* beneficial to the natural world? Both positions encounter ethical and policy-oriented problems. The anthropocentric perspective would permit the use (and destruction) of natural entities for a correspondingly greater human benefit; but the nonanthropocentric intrinsic value perspective implies a policy of strict nonintervention in natural processes, an absolute sanctity of nature. One position may lead to the destruction of nature, and the other may lead to worshipful noninterference: thus the dilemma for environmental philosophers.

On a practical level, the theocentrism of Judaism resolves this dilemma because it is functionally equivalent to a nonanthropocentric doctrine of the intrinsic value of nature without endorsing the sacredness of natural entities in themselves. Natural objects are valued, and cannot be destroyed, because they belong to God. They are sacred, not in themselves, but because of God's creative process. This worldview is, in part, derived from the Kabbalistic strand of Jewish thought, as is expressed quite clearly by David Shapiro: "The quality of lovingkindness is the basis of all creation. It is God's steadfast love that brought this world into being, and it is His steadfast love that maintains it."[44] Thus, "all creation is linked together in a bond of unity," which humans must act to preserve and not to destroy.[45] A further description of this view is offered by Lamm, who writes that "Judaism . . . refuses to ascribe the quality of holiness to nature and natural objects as such."[46] The Jewish view of the human relationship to nature can be represented by the opposition of two extreme views, with the mainstream Jewish tradition taking the middle position. On the one side is the form of Hasidism that follows the Kabbalistic tradition of God's immanence throughout nature, the extreme of nature-deification. On the other side is the Mitnagdic criticism of Hasidism which radically separates the divine and natural realms, the extreme of nature-as-profane. Lamm argues that the two extremes tend to converge, for the Hasidic tradition teaches respect for nature without ascribing sanctity to it, while the Mitnagdic tradition acknowledges that from God's perspective the world is suffused with his presence. For Lamm, this "theological tension is resolved . . . [in] . . . practice . . . [as] Nature is not to be considered holy, but neither is one permitted to act ruthlessly towards it, needlessly to ravage it and disturb its integrity."[47] As Gordis concludes, "every natural object is an embodiment of the creative power of God and is therefore sacred."[48] Its sacredness and its integrity—its intrinsic value, let us say—rests on its status as God's creation. Thus, it is the theocentric basis of *bal tashchit* that requires Jews to act with a practical respect for the value of nature without regard to human concerns.

The Jewish View of Nature

This survey of Jewish principles and commandments regarding nature and the environment does not lead easily into a unified worldview. Is it possible to summarize this examination of the specific regulations of Judaism concerning nature? Is there a coherent Jewish perspective? Yes: the Jewish worldview holds that nature has a value independent of human interests, as an expression of the creative power of God. This divinely inspired value thus inspires respect and requires obedience on the part of humanity, the servants and stewards of God's creation.

As stewards of God's earth, humans serve as partners in the never-ending task of perfecting the universe. Gordis concludes that "Judaism . . . insists that human beings have an obligation not only to conserve the world of nature, but to enhance it" as a "copartner with God in the work of creation."[49] The universe is God's creation, and that is the undeniable and fundamental starting point of the Jewish view of nature. Understanding the universe as an outgrowth of God's power is the most important aspect of the value of nature in the Jewish worldview. It gives the natural world a force, a presence, that cannot be ignored.

Allen ends his discussion of the value of nature with a return to the book of Job, for in God's speech out of the whirlwind we are presented with the essence of the wild: a world beyond the control and understanding of humanity. But the lack of control does not breed disrespect; on the contrary, it creates a sense of awe, wonder, and responsibility, for we are in the presence of the divine. "The untamed world beyond the frontiers of human society is fraught with the numinous, it is a constant reminder that man is not master in the world but only a privileged and therefore responsible inhabitant of it."[50]

Notes

1. Robert Gordis, "Judaism and the Environment," *Congress Monthly* 57, no. 6 (September/October 1990): 8. This article is a revised version of "Judaism and the Spoilation of Nature," which appeared in *Congress Bi-Weekly*, 2 April 1971. Another version of the essay, with substantial similarities, appeared as "Ecology and the Jewish Tradition" in *Judaic Ethics for a Lawless World* (New York: Jewish Theological Seminary, 1986), 113–22. This 1986 essay is reprinted in *Judaism and Ecology, 1970–1986: A Sourcebook of Readings*, ed. Marc Swetlitz (Wyncote, Pa.: Shomrei Adamah, 1990), 47–52.

2. Jeanne Kay, "Concepts of Nature in the Hebrew Bible," *Environmental Ethics* 10 (1988): 326–27.

3. E. L. Allen, "The Hebrew View of Nature," *The Journal of Jewish Studies* 2, no. 2 (1951): 100.

4. Ibid.

5. Lynn White, Jr., "The Historical Roots of Our Ecologic Crisis," *Science*, no. 155 (1967): 1203–7.

6. For a full philosophical discussion, see John Passmore, *Man's Responsibility for Nature: Ecological Problems and Western Traditions* (New York: Scribner's, 1974), 3–40, and Robin Attfield, *The Ethics of Environmental Concern* (New York: Columbia University Press, 1983), 20–87.

7. Norman Lamm, "Ecology and Jewish Law and Theology," in *Faith and Doubt* (New York: KTAV, 1971), 164–65; reprinted in Swetlitz, *Judaism and Ecology*, 77.

8. See Gordis, "Judaism and the Environment," 7–8.

9. David Ehrenfeld and Philip J. Bentley, "Judaism and the Practice of Stewardship," *Judaism* 34 (1985): 301–2; reprinted in Swetlitz, *Judaism and Ecology*, 97–98.

10. Ibid., 305 (in Swetlitz, 99).

11. Jonathan I. Helfand, "The Earth Is the Lord's: Judaism and Environmental Ethics," in *Religion and Environmental Crisis*, ed. Eugene C. Hargrove (Athens: University of Georgia Press, 1986), 39.

12. Ibid., 40.

13. Samuel Belkin, "Man as Temporary Tenant," in *Judaism and Human Rights*, ed. Milton R. Konvitz (New York: Norton, 1972), 253; reprinted in Swetlitz, *Judaism and Ecology*, 25.

14. Ibid., 253–54 (in Swetlitz, 25–26).

15. Ibid., 255 (in Swetlitz, 26).

16. Helfand, "The Earth Is the Lord's," 40–41.

17. Belkin, "Man as Temporary Tenant," 252 (in Swetlitz, 25).

18. Ehrenfeld and Bentley, "Judaism and Stewardship," 309 (in Swetlitz, 101).

19. Ibid., 310 (in Swetlitz, 102).

20. Ibid., 306–7 (in Swetlitz, 100).

21. See discussion in Eric G. Freudenstein, "Ecology and the Jewish Tradition," *Judaism* 19 (1970): 409–10; reprinted in Swetlitz, *Judaism and Ecology*, 30–31. See also Richard H. Schwartz, *Judaism and Global Survival* (New York: Vantage Press, 1984), 49.

22. Helfand, "The Earth Is the Lord's," 46. See also Aryeh Carmell, "Judaism and the Quality of the Environment," in *Challenge: Torah Views on Science and Its Problems*, ed. Aryeh Carmell and Cyril Domb (New York: Feldheim, 1978), 511; reprinted in Swetlitz, *Judaism and Ecology*, 39.

23. Helfand, "The Earth Is the Lord's," 46.

24. Carmell, "Judaism and the Quality of the Environment," 500–525 (in Swetlitz, 34–46).

25. Ibid., 503 (in Swetlitz, 35).

26. Ibid., 504 (in Swetlitz, 36). Carmell cites the Mishna, *Bava Bathra* 2:8–9.

27. Ibid., 505 (in Swetlitz, 36).

28. Helfand, "The Earth Is the Lord's," 42.

29. Ibid., 45.

30. Gordis, "Judaism and the Environment," 8.

31. Allen explains that Judaism is not a Schweitzerian ethic, an ethic for all life, because there is a difference between domestic and wild animals. Humans have a specific covenant with domestic animals to protect them; the concern for wild nature is more mysterious, as it is based on the recognition of the divine presence in the entire world of creation. This point will be expanded below. See Allen, "Hebrew View of Nature," 103.

32. Gordis, "Judaism and the Environment," 9.

33. Lamm, "Ecology and Jewish Law," 169 (in Swetlitz, 79).

34. Ibid.

35. Lamm, 169 (in Swetlitz, 79) and Gordis, 9.

36. Gordis, ibid., cites *Sifre Shofetim*, section 203.

37. Lamm, 169 (in Swetlitz, 79).

38. Gordis, 9.

39. Lamm, 170 (in Swetlitz, 80). Lamm cites *Turei Zahov* to *SH.A.Y.D.* 116:6, and Responsa *Havot Yair*, no. 195.

40. Freudenstein, "Ecology and Jewish Tradition," 411 (in Swetlitz, 31).

41. Ibid., 411–12 (in Swetlitz, 31–32).

42. Lamm, 171–72 (in Swetlitz, 80–81). See also Gordis, 9.

43. Gordis, 9.

44. David S. Shapiro, "God, Man and Creation," *Tradition* 15 (1975): 25; reprinted in Swetlitz, *Judaism and Ecology*, 64.

45. Ibid., 41 (in Swetlitz, 73).

46. Lamm, 173 (in Swetlitz, 81).

47. Ibid., 173–77 (in Swetlitz, 81–83).

48. Gordis, 10.

49. Ibid., citing *B. Shabbat* 10a.

50. Allen, "Hebrew View of Nature," 103. I have discussed these themes from a secular standpoint in Eric Katz, "The Call of the Wild: The Struggle against Domination and the Technological Fix of Nature," *Environmental Ethics* 14 (1992): 265–73.

The Garden of Eden, The Fall, and Life in Christ:
A Christian Approach to Ecology

Jay McDaniel

Hendrix College

HISTORICALLY, the Western Christian approach to nature has been thoroughly ambiguous. While it is an exaggeration to claim that Christianity is largely responsible for negative attitudes toward nature, it is also an exaggeration to claim that it is largely innocent. The truth lies in between.

The ambiguity is displayed in H. Paul Santmire's *The Travail of Nature: The Ambiguous Ecological Promise of Christian Theology*.[1] Santmire suggests that two motifs have woven their way through Western theology: (1) a "spiritual" motif, in which the end of human existence is thought to lie either in a transcendence of nature or, in modern times, a humanization of nature; and (2) an "ecological" motif, in which the end is thought to lie in community with nature, appreciative of nature's blessings and cognizant that nature has value apart from its usefulness to humans.

As Santmire sees it, representatives of the spiritual motif include Origen (ca. 185–254), Thomas Aquinas (1225–74), Bonaventure (1221–74), Dante (1265–1321), Karl Barth (1886–1968), and Teilhard de Chardin (1881–1955). Despite the fact that some of these theologians recognized the splendid fecundity of the earth, their appreciation was obstructed by an overly spiritual conception of the end of human life. Origen, for example, saw the goal of human life as a return of the fallen rational spirits to perfect union with God, in which the rest of creation is left behind (*TN*, 44–53). Thomas, after having affirmed the goodness of creation in a way that Origen never did, nevertheless claimed that the end of the whole creation is chiefly the beatitude of human beings (and the angels) alone in heaven with God (*TN*, 85–95). Dante and Bonaventure recognized the fecundity of nature, but they viewed nature primarily as a ladder to heaven which was no longer

71

needed once heaven was attained (*TN*, 97–106). For Barth, nature was a divine afterthought that allowed God's primary purpose—redemption of human beings—to be fulfilled (*TN*, 145–55). And for Teilhard, nature was a stepping-stone to humanity, an evolutionary process to be humanized through technological progress (*TN*, 170).

Of course, in the case of Teilhard, the very emphasis on evolution and its inner nature is helpful for ecologically minded thinkers, Christian and otherwise. A Teilhardian approach to ecology is eminently possible. Indeed, the neo-Teilhardian cosmological approach of Thomas Berry and Brian Swimme is represented in this volume. Still, Santmire's critique is that, for Teilhard, the whole point of nature is to make way for humanity. Here, as in Barth, nature is not fully valued for its own sake; it is valued for human's sake. In this respect, according to Santmire, Teilhard illustrates the spiritual motif that finds human fulfillment in a transcendence or transformation of nature.

Representatives of the ecological motif include Irenaeus (ca. 130–200), the later Augustine (354–430), Francis of Assisi (1182–1226), and, to a lesser degree, Luther (1483–1546) and Calvin (1509–64). With his vision of a good creation that will be redeemed as a whole at the end of time, "Irenaeus celebrates the flesh in this world and in the life to come . . . Nature for him is tangibly good and ultimately significant" (*TN*, 176). Similarly, the mature Augustine sees that "all things, the creatures of nature as well as human creatures, have their own integrity, their own value, their own necessary place in the greater history of the created order" (*TN*, 177). And in Francis, says Santmire, we find a fulfillment of the ecological promise of the classical Christian ethos: "Francis climbs the mountain of God's creation in order to stand in universal solidarity with all God's creatures, both in this world and the world to come" (*TN*, 178).

Even Luther and Calvin, not generally known for their sensitivity to nature, have promising aspects for Santmire. Both reject the spiritual motif of ascent to God, emphasizing instead that God "descends" into the world. At points, both speak of grace through nature and solidarity with nature. But their general attitude toward nature is much more ambivalent than that of Irenaeus, the mature Augustine, and Francis. They see nature primarily as a means to human ends and a stimulus to the human reception of grace. Luther, for example, stresses that nature is oftentimes a hostile stimulus. Moreover, both Luther and Calvin are so preoccupied with issues pertaining to human salvation, such that themes of grace through nature and solidarity with nature remain at the margins of their thought (TN, 179–89). Compared to Irenaeus, the later Augustine, and Francis, they only partially embody the ecological motif. After tracing the two motifs, Santmire observes that the anthropocentrism implicit in the spiritual motif was secularized in the modern era, fueled by the rise of modern science and the industrial revolution. He concludes that there is a profound ambiguity in Christian attitudes toward nature and that the negative side of this ambiguity reached prominence in the modern period (TN, 121–43).

The question then emerges: and what about the future? Today increasing numbers of Christian theologians are developing creation-centered forms of theology.[2] Many recognize that the Christian heritage, even with its positive contributions to ecological sensibilities, is not entirely adequate to an ecological age. They see a need to extend the Christian heritage (1) by accentuating the positive contributions of the heritage to ecological concerns; (2) by critiquing the negative; and (3) by opening themselves to new insights drawn from the natural sciences, from a dialogue with other world religions, and from the earth itself.[3] They see Christianity itself not as a timeless set of creeds but rather as an ongoing tradition capable of self-criticism and improvement, repentance and new life.[4]

There are many types of creation-centered theology now being developed. Some are feminist and some biblical, some are first world and some third world, some are mystical and some political, some cosmological and some existential, some Catholic and some Protestant. In what follows I offer what might be called a "liberal Protestant" approach. Liberal Protestants are those who take the Bible seriously as a source for wisdom, but who also see reason, tradition, and experience as sources of wisdom. In the spirit of liberal Protestantism, I offer in what follows a creation-centered reading of three biblical themes: Dominion, the Fall, and Life in Christ.

Dominion

In the Bible "dominion" (rada) generally means "rule" or "lordship." As is well known, the first creation story in Genesis 1:26 says that humans are to have dominion over the earth. Biblical scholars disagree on how such dominion is to be interpreted. Gerhard Von Rad believes that dominion in Genesis has dominating overtones. He notes that *rada* itself has connotations of trampling, as when one tramples a wine press.[5] By contrast, James Barr believes that dominion refers to the kind of beneficent governance that, for example, Solomon exercised over his realm.[6] On Barr's reading, the dominion that humans have over the earth is to be analogous to the governance of a kindly ruler who seeks to protect and preserve a cherished realm. For Barr, dominion means stewardship.

Of course, even *if* we assume that dominion means stewardship, the very idea of stewardship can be problematic. The idea easily lends itself to attitudes of separation from the rest of creation. If dominion-as-stewardship is to be affirmed, emphasis must be placed on the fact that the stewards themselves are creatures among creatures, human nodes in the broader web of life.

Interestingly, the first creation story does indeed situate human beings in a broader cosmic context. To be sure, the story stresses that humans, and humans alone, are made in God's image. But the notion of *imago dei* does not alter the fact, from the point of view of the biblical narrative, that humans are creatures among creatures, created by God along with the other animals on the sixth day. At the end of the narrative, it is the whole of creation, not humans alone, that is declared "very good." Thus, *imago dei* in no way denies human participation in the web of life or human mortality; nor does it deny the sheer goodness of creation apart from its usefulness to humans. From a perspective sensitive to ecological concerns, *imago dei* refers to the unique potential to care for creation with a stewardly compassion that mirrors God's own.[7]

Still, the question remains: should Christians retain the idea of dominion at all? I believe that they should, because "dominion" names a historical fact. Today, humans *do* have a certain kind of rule over the rest of the earth. Our dominion is our powerful influence over plants, animals and their habitats, both locally and in modern times globally. We exercise it daily through agriculture, urban development, resource extraction, and other forms of manipulation characteristic of modern, industrial civilizations.

Furthermore, our dominion is now irreversible. By United Nations estimates, the human population will be around 6.3 billion by A.D. 2000 and 11 billion by the end of the next century.[8] It would be impossible for 6.3 billion people to live on the planet without exercising inordinate rule over other creatures and their habitats, if only to meet basic needs for food and shelter. To meet these needs, much manipulation will be required, for good or ill.

The only alternative to such massive manipulation would be a reduction of the human population to preindustrial levels of five hundred million or less. Of course, a reduction of this sort could result from some combination of nuclear war, disease, famine, and ecological collapse. But most of us wish to avoid such calamities, and for good reasons. A violent reduction of this sort would involve unimaginable tragedy not only for most humans—whose lives, like those of other creatures, are precious—but also for members of other species, who would perish through habitat destruction and other forms of violence. To avoid such catastrophe, our best option is (1) to accept the ambiguity of such a high number of humans on the planet; (2) to stabilize that population as much as possible; and then (3) to find ways of allowing six to eleven billion people to live on the planet in ways that are ecologically wise. In the best of scenarios, we are doomed to dominion.

As we exercise our dominion, we need an image of what might be called *right dominion*. Here the idea of dominion-as-stewardship can help. It offers an image of the right way to exercise power in relation to nature, as opposed to the wrong way upon which so many industrial civilizations are now embarked. At least this is the case if stewardship itself is understood to involve wise management undergirded by respect for life and environment. This, I believe, is what the biblical idea of stewardship recommends at its best. Understood as kindly use in a spirit of respect, the image of stewardly dominion invites us to maximize the quality, not the quantity, of human life, with minimum abuse of domesticated animals and minimum impact on wildlife and habitats. It invites us to develop societies that are ecologically sustainable even as they are socially just.[9]

The Garden and the Fall

Still, there must be more to a contemporary Christian approach to ecology than an accent on stewardly dominion. Somehow, dominion itself must be understood in terms of a larger view of human history. This takes us to the theme of the Fall and its precursor, the Garden of Eden.

Christians have traditionally treated the age of the Garden as that time in human history when humans lived in relative harmony with one another and the rest of creation. It was an age of innocence. I submit that the closest approximation we have to this mythical image is itself the longest and most enduring period of human history: the age of the hunters and gatherers.

For more than two and a half million years, human rule over other creatures and their habitats was minimal compared to our own, and it was for the most part benign. During this period the early hominids and the first humans lived in families of twenty to forty, foraging for food and hunting game. Archeological evidence shows that they lived well by foraging and hunting; that they buried their ancestors in special ways; that they knew the plants and animals in their immediate environments with intimacy; and that, as cave paintings make clear, they had distinctive creative talents. Moreover, they seem also to have had a spirituality of sorts. At least this is what the philosopher Max Oeschlaeger suggests.

Synthesizing data from contemporary anthropology and paleoanthropology, Oelschlaeger identifies several feelings that, in his view, were part and parcel of the ancient hunting-gathering mentality. As he sees it, our earliest ancestors felt (1) that irrespective of place, nature was home; (2) that nature was alive; (3) that the entire world of plants and animals, even the land itself, was sacred; and (4) that divinity could take many natural forms.[10] If Oeschlaeger is correct, then during this period, our earliest ancestors felt no special separation from the rest of nature. They were in the web of life, not apart from it. Like Adam and Eve before the Fall, they were relatively innocent. They did not know they were naked.

The Fall

When, then, did the Fall occur? When did our powers of dominion increase and our potential for domination emerge? Some might point to the emergence of modern, industrial civilizations in the seventeenth and eighteenth centuries. Certainly human dominion increased rapidly during this period, often in destructive ways. But the better answer, I submit, is some twelve thousand years ago, with the dawn of plow agriculture. Wes Jackson, an agroecologist and interpreter of the Christian tradition, puts it this way: "The Fall . . . can be understood in a modern sense as an event that moved us from our original hunting-gathering state, in which nature provided for us exclusively, to an agricultural state, in which we took a larger measure of control over food production, changing the face of the earth along the way . . . I suspect that agriculture is at the core of the Fall."[11]

In history books, of course, we do not speak of the emergence of agriculture-based civilizations as a fall. Rather we speak of it as a collective advance for the human species. We celebrate the tremendous creativity involved in spinning and weaving, brick making and mortaring, mining and smelting, law and religion, in urban life itself. Most of us are not willing to give up these benefits. With the Christian tradition, we recognize that the Fall itself was ambiguous, because it brought with it advantages as well as liabilities.

Still, the Fall did indeed have liabilities. It brought with it many human evils unknown to the early foragers, including slavery, patriarchy, organized warfare, and a loss of ecological innocence. With the rise of farming, fields and settlements began to displace animal habitats at alarming rates. As food sources became more stable, the human population began to increase dramatically, thus requiring still more habitat destruction. In many ways humans became not plain citizens of the planet but lords of the planet. Much of our dominion became domination.

The Psychology of the Fall

J. Baird Callicott emphasizes still another, more psychological, dimension of the Fall (*CNC*, 122–24). According to Callicott, the Fall involved the gradual emergence of a certain mode of consciousness which is both a blessing and a curse in human existence. Genesis 3:5 speaks of it as "the knowledge of good and evil."

Many of us are socialized into this knowledge by our cultures. We almost consider it essential to human existence. Two aspects of the knowledge are worth noting. First, the knowledge involves a sense of self-awareness, such that we understand ourselves as separate from other creatures in certain ways, and as capable of exercising inordinate rule over them and, for that matter, over one another. With such capacities comes a capacity to recognize that, after all, we have choices to make in whether and how we will exercise our newfound power. We become aware that there are good and bad ways to use power. We have "the knowledge of good and evil."

Second, "the knowledge of good and evil" involves a kind of ranking, or evaluating, of items in the world as "good" or "evil" relative to their human utility. Of course, hunters and gatherers partook of such ranking. They knew that some plants were edible and others not. But in the urban-agricultural civilizations this ranking is intensified. Some plants become "good" plants and

others become "weeds"; some animals become "good" animals
and others become "varmints." With the Fall into agriculture, so
Callicott suggests, anthropocentrism emerged in many cultures.
Callicott suggests that anthropocentrism was, and is, the original
sin (CNC, 125).

Life in Christ

In light of this original sin, what are humans to do? Can we
return to the Garden? The Bible says no, because angels block
the way. Ecological wisdom tells us also that the earth blocks the
way. Hunting and gathering societies require considerable territory to support a relatively small number of individuals.[12] The
planet cannot hold six billion hunters and gatherers.

After the Fall, we can only move forward into as yet uncharted,
but hopefully benign, relations with the rest of creation. Callicott
puts the point this way: "We can no more re-create the original
Garden of Eden than we can recover our original unself-con-
sciousness. But we can try to live harmoniously in and with nature
. . . by employing all of our postindustrial ingenuity and ecological
understanding to create an environmentally benign sustainable
civilization" (CNC, 132).

If our civilizations are to be sustainable, we need to become the
kind of people, with the kinds of sensibilities, that can sustain
such communities once they emerge. It is at this point, I believe,
that the inner dimensions of the Christian life become relevant.

Christian spirituality does not offer a way of returning to the
Garden of Eden, however understood. It does not promise a recovery of lost innocence or a return to "original unself-conscious-
ness." Instead, it encourages us (1) to accept our lost innocence;
(2) to trust that, despite the loss, we are embraced by an all-com-
passionate Mystery—a Christlike God—who seeks our own good
and that of the whole creation; and (3) to open our hearts to the
healing powers of this Mystery so that we become vessels of the
Mystery's own love. This is the emphasis of Paul in the New Testament, most notably in his letter to the Romans. However the nuances of the epistle are interpreted, the epistle as a whole points
to a way of living in which humans are grasped by a healing grace
that is given freely by God to people who have lost their innocence.

Indeed Pauline spirituality goes a step further. It suggests that
the God who bestows such healing grace can be immanent within

our own lives. To the degree that we are open to these healing powers, the divine Mystery unfolds within us, as a self deeper than our own egos, who lives through us yet is more than us. As present within us, God is the living Christ. To live from nourishment of the living Christ is to enjoy life in Christ. We are able to say, with Paul: "It is no longer I who live, but Christ in me" (Gal. 2:20).

The living Christ can be conceived in different ways by different Christians. Conservatives are inclined to conceive the living Christ as a divine reality who enters us after we have accepted our lost innocence and accepted God's grace. Accepting God's grace is itself tied to accepting Jesus as personal lord and savior. Thus, for conservatives, "life in Christ" is limited to born-again Christians.

More liberal Christians have an alternative view. They see "the living Christ" as a divine consciousness that is already within us at a deep level, even prior to our recognition of it, to which we awaken. Here "Christ" does not enter from afar, rather "Christ wells up from within, to the degree that we are in touch with the deepest center of our lives.

Awakening to the living Christ is itself an evolutionary possibility, the way for which was paved by billions of years of cosmic evolution and millions of years of biological evolution. Indeed, this possibility became available and relevant only after the Fall into agriculture. Our distant elders, the early foragers, had their own ways of dwelling in communion with God, relative to their time and situation, as do other animals. Life in Christ is for people who have lost their innocence. It is one among many modes of religious existence available after the urban-agricultural revolution.

Historically, the significance of life in Christ can be understood in two ways. On the one hand, it can be understood as a postagricultural salvage operation, a way of making the best of a bad situation. Understood in this way, "life in Christ" is a way of enjoying redemption, that is, of enjoying divine forgiveness despite human sinfulness.

On the other hand, "life in Christ" can also be conceived as an advance in the history of life on earth, at least if it is lived to its fullest. Understood in this way, it is more than a way of being saved from the consequences of the Fall. It is a contribution to the very history of creation, without which creation would not be as rich.

To conceive of "life in Christ" as contributing to the history of creation, of course, is to believe that human history, and the universe itself, is an ongoing process capable of growth and change and that new things can happen which lack precedence in the past. This is a traditional Christian view, rooted in the prophetic idea that God can offer new possibilities in new historical situations. It is the idea that new possibilities emerge from the Life at the heart of the universe, that were not available or relevant beforehand, and that offer genuine possibilities for "new life" relative to the situations at hand. "Life in Christ" is a kind of new life.

Among Christians, what is sorely needed in our time is a spelling out of some of the implications of this new life for ecology. Recall that "life in Christ" involves (1) an acceptance of lost innocence; (2) a recognition of the limitless love of God; and (3) an openness to the healing powers of this God as they well up from within the very depths of our existence. To accept our lost innocence is to accept the fact that many of us in urban-industrial civilizations *do* feel separate from one another and the rest of creation and that, in the latter regard, we *do* tend to evaluate other creatures in light of their usefulness to us. We must confess that we partake of "anthropocentric consciousness." This is part of our sinful existence.

To recognize that we, and the whole of creation besides, are embraced by a Christlike Mystery is then to understand that this "fallen consciousness" is not the ultimate perspective. It is to realize that the most inclusive perspective belongs not to us but to God, who loves each and every creature for its own sake: the amoeba no less than the chimpanzee, the mosquito no less than the human. This does not mean that God loves all creatures equally or in the same way. There may be more to love in the chimpanzee than the amoeba, by virtue of the richness of the chimpanzee's sentience. The point is simply that each creature is loved by God on its own terms and for its own sake, however rich those "terms" might be. The Protestant theologian Schubert Ogden puts the point this way: "Because God's love itself is subject to no bounds and excludes nothing from its embrace, there is no creature's interest that is not also God's interest and, therefore, necessarily included in the redeeming love of God."[13]

To be healed by God is then to allow one's life, as best one can, to be a vessel for this limitless love. I have proposed that, from a Christian perspective, the limitless love wells up from within the depths of our existence, as a "living Christ" who is our deepest center. To the extent that we partake of this love, our own "dominion" will be tempered by a deeper recognition of the sheer good-ness, the sheer lovability, of each and every living being whom we influence. As Paul puts it, we will have put on the mind of Christ (Phil. 2:5).

Of course, putting on the mind of Christ is no mean feat. It entails worship and prayer, study and service, community and contemplation, all of which are best understood as ways of cooper-ating with divine grace. Moreover, God-consciousness occurs in degrees. Few Christians have put on the mind of God fully. Some in other religions have done so better, without even using the word "Christ." In any case, "life in Christ" is an ideal to be ap-proximated by finite creatures. To the degree that Christians learn to feel the world as God feels it, with sensitivity to the intrin-sic value of each and every life, and with delight in the sheer diversity of forms of life, we will approximate "life in Christ." And to the degree that we approximate this life, we will be the kind of people the world sorely needs today. Such is a Christian ap-proach to ecology.

Notes

1. H. Paul Santmire, *The Travail of Nature: The Ambiguous Ecological Promise of Christian Theology* (Philadelphia: Fortress Press, 1985); hereafter, *TN,* with page references cited in the text.

2. See *Liberating Life: Contemporary Approaches to Ecological Theology,* ed. William Eakin, Charles Birch, and Jay McDaniel (Mary Knoll, N.Y.: Orbis Books, 1990).

3. For the role of the natural sciences in contemporary ecological theologies, see Charles Birch, "Chance, Purpose, and the Order of Nature" and also "Liberating Life: A Report to the World Council of Churches," both in *Liberating Life.*

4. For examples of new vision, see Thomas Berry, "The Spirituality of the Earth"; John F. Haught, "Religion and Cosmic Homelessness: Some Environmental Implications"; Sallie McFague, "Imaging a Theology of Nature: The World as God's Body"; and Jay McDaniel, "Revisioning God and the Self: Lessons from Buddhism," all in *Liberating Life.*

5. Gerhard Von Rad, *Genesis: A Commentary* (Philadelphia: Westminster Press, 1961), 60. Cited in Carol Adams, "Feeding on Grace," in *Good News for Animals: Christian Ap-proaches to Animal Well-Being,* ed. Charles Pinches and Jay McDaniel (Mary Knoll, N.Y.: Orbis Books, 1993), 155.

6. James Barr, "Man and Nature: The Ecological Controversy and the Old Testament," in *Bulletin of the John Rylands University Library of Manchester,* 9–32. Cited in Adams, "Feed-ing on Grace," 155.

7. See Jay B. McDaniel, *Earth, Sky, Gods, and Mortals: Developing an Ecological Spirituality* (Mystic, Conn.: Twenty-Third Publications, 1990), 165.

8. Jerome Fellmann, Arthur Getis, and Judith Getis, *Human Geography: Landscapes of Human Activities* (Dubuque, Iowa: Macmillan, 1985), 96.

9. The New Testament offers a theme that can further sensitize us to an understanding

of right dominion. From the vantage point of the Gospels, authentic human rule—at least over other humans—is a reversal of our usual expectations. It is the rule of nonviolent service offered in a spirit of self-sacrificial love: "The Son of Man came not to be served but to serve, and to give his life as a ransom for many" (Matt. 20:28). In the history of Christianity, the peace traditions—Mennonites, Hutterites, and Quakers—have taken this reversal of "rule" imagery most seriously. They have seen the Christian life as one of service, even to the point of refusing to kill. In an ecological age, Christians can follow their lead and extend it. They can recognize that nonviolence extends also to our treatment of animals and the earth, and that in many ways our task is to serve the whole of life, not be served by it. Such would be the nature of a Christlike dominion.

10. Max Oelschlaeger, *The Idea of Wilderness* (New Haven: Yale University Press, 1991), 12.

11. Quoted in J. Baird Callicott, "Genesis and John Muir," *Covenant for a New Creation: Ethics, Religion, and Public Policy,* ed. Carol S. Rob and Carl J. Casebolt (Mary Knoll, N.Y.: Orbis Books, 1991), 125; hereafter, *CNC,* with page references in the text.

12. Fellmann, *Human Geography,* 36.

13. Quoted from the frontispiece of *Liberating Life.*

The Ecological Fallout of Islamic Creation Theology

Roger E. Timm

Carthage College

DEBATE about the influence of religious beliefs on attitudes toward the environment has often focused on the implications of the creation narratives in the Bible. Scholars concerned about environmental issues have charged that these accounts have been used to support exploitation of the earth and its resources.[1] If one agrees with this thesis, it may be only natural to assume a corollary assertion that any monotheistic creation theology would lead to similar exploitative approaches to the natural environment. Since Muslims clearly believe in one God who is the Creator of the universe, examining the "ecological fallout" of the Islamic worldview may be a helpful way of testing this corollary assertion. Specifically, in this essay I would like to address the question, "What attitudes toward the earth and its resources are implied by the belief in Allah as Creator in the Islamic religious tradition?"

Later I will make some comments about how Muslims past and present have treated the environment. I am interested first, however, in exploring the *conceptual* implications of the early Islamic tradition.[2] That is, are there materials in the authoritative literature of the early Islamic religious tradition that could justify exploitation of the environment similarly to the way some believe the creation accounts in Genesis do? Since the Islamic religious tradition shares a common place of origin in the Middle East with the Jewish and Christian traditions, one might suppose that Islam does understand the meaning of creation in a similar way. Is this in fact the case? The answer turns out to be both "yes" and "no."

Creation in Early Islam

The references to Allah's creating the world in the Qur'an are clearly related to the creation accounts in the Bible, but the differences between them are striking.[3] The Qur'an includes no single creation story as complete as the accounts in Genesis; rather it contains repeated references to elements of several creation stories that perhaps were commonly known.[4] Like the Bible, the Qur'an states that Allah created the world in six days (7:54; 10:3; 11:7; 25:59; 32:4; and 57:4),[5] but the assertion that God rested on the seventh day is explicitly rejected, because it seems to demean the glory and power of God to suggest that rest was needed after the "work" of creation (50:38).[6]

Although the Qur'an has no detailed account of the creation of the first human, it does agree with the Bible in naming the first human Adam. The Qur'an, like Genesis 2, describes Adam as being created from dust (3:59; 35:11; and 40:67) or, as suggested by the biblical image of God as a potter, out of clay (6:2; 15:26; 32:7; 38:71; and 55:14; cf. *Mishkat*, 33). Like the Bible, the Qur'an teaches that Adam gave names to what Allah created, but, unlike the Bible, it describes Adam as merely expressing what God had revealed to him, thereby demonstrating God's power and knowledge, not Adam's (2:31–33). The Qur'an clearly indicates that the first humans were created a male-female pair, but the first woman is referred to as Adam's wife and does not seem to be known as "Eve" (2:35 and 4:1).[7] Another difference from the Bible is that the Qur'an denies that humans were created in the image of God, apparently lest Allah's majesty be demeaned (42:11).

The God of Creation

The distinctiveness of the Qur'an's description of creation becomes especially clear when its theological emphases are considered. References to creation, for example, serve primarily to focus on Allah and various attributes of Allah.

Perhaps Islam's most central affirmation is that "there is no God but Allah." Repeatedly in the Qur'an references to creation underscore this basic affirmation of the oneness of God. Those who ascribe partners or rivals to God are condemned (6:1; 10:34; 13:16; 27:59–61; and 40:62).[8] Some of these passages seem to be criticizing Muhammad's Arab contemporaries for their polytheism, believing in "rivals" to Allah and even in daughters of Allah, and his Christian neighbors for their belief in the divinity of Jesus, assigning a "partner," or a son, to Allah (6:95–103; 16:20; 22:73; 25:2–3; 35:40; and 37:149; and al-Bukhari, vol. 4, bk. 54, ch. 1, 279; and vol. 6, bk. 60, ch. 10, 11, and ch. 357,

470).[9] The Qur'an finds such beliefs without sense or merit; it expresses the view that nature and reason clearly witness to the unity of Allah, the one Creator of all.

The Qur'an emphasizes not only the oneness of Allah but also Allah's sovereignty and power over all creation. Allah "has power over all things" is a frequent refrainlike phrase in the Qur'an (2:106 and 109; 3:29; 5:19; 8:41; 9:39; 22:6; 29:20; 46:33; and 64:1; cf. Bell, 148). The power of Allah, however, is not only potential but also actual. Another commonly recurring phrase states that the dominion over the heavens and the earth belongs to Allah (2:107; 3:189; 9:116; 23:84–89; 35:13; and 57:2).[10] Nothing escapes God's power and authority, and all things in heaven and earth are Allah's (2:116 and 255; 10:55; 16:52; 30:26; and 53:31). God's power is so great that God can create at will. All that Allah needs to do is to say, "'Be' and it is" (2:117; and 42:49).

While Allah is almighty, the Qur'an also depicts God as beneficent and merciful. God's creation is a gift of God's grace (28:73; 35:3; and 53:32). I believe that it is important to emphasize this, for some commentators on Islam have suggested that its affirmation of the divine power of God leads to a view of Allah as stern and intimidating. To suggest this, however, is to misrepresent the customary Islamic view of God. The *hadith* literature, for example, reports Muhammad as saying, "When God completed the creation He wrote the following which is with Him above His Throne, 'My mercy has taken precedence over my anger'" (al-Bukhari, vol. 4, bk. 54, ch. 1, 279; and *Mishkat*, 502).[11]

In the Islamic tradition both the sovereignty and the beneficence of Allah are described as present realities, indicating that the creative power of God continues to be expressed in Allah's ongoing creation of the world. Allah is the sustainer and guider of all life, in the present as well as the past (13:17; 15:16–23; 20:50; 30:40; 43:11; and 56:63–74).[12] There is a sense in which it may be said that Allah continually recreates the world in every moment. Or as the Qur'an puts it, "God originates creation, then repeats it" (10:4; 29:19–20; 30:11 and 27; and 85:13).[13] The creation theology of early Islam, then, is not intended primarily to describe some primordial act of God's creativity but rather to confess Allah's ongoing creative role in the present life of believers and their world.

Moreover, one may well argue that the creation theology of early Islam points to God's activity primarily at the conclusion of the world process, for the Qur'an also describes Allah as a judge.

The Qur'an says, "He it is Who created the heavens and the earth in six Days ... that He might try you, which of you is best in conduct" (11:7).[14] Allah, the originator of life, is also the end of the journey; "to God is the final goal" (24:42; cf. 18:48 and 64:3).[15] In fact, many of the Qur'an's references to the creation are part of arguments intended to show Allah's ability to raise the dead for the final judgment (17:51 and 98–99; 19:66–67; 22:5; 46:33; and 50:15).[16] Even an analysis of the creation accounts in the Qur'an, then, demonstrates Islam's eschatological flavor.

The Purpose of Creation

Allah, the Qur'an says, created the world neither "in jest" nor "for naught"—that is, God had a serious purpose for the creation (3:191; 21:16; 23:115; 38:27; and 44:38–39).[17] What is that purpose? Several answers are given in the early Islamic tradition. The foremost purpose of creation is to serve as a collection of "signs" (ayat) of the power and goodness of Allah. People have no excuse for not believing in God, for the world is filled with signs of Allah's creative activity (2:164; 3:190; 13:2–4; 16:10–13; 27:86; 29:44; 30:20–25; 41:37; 45:3–6; and 88:17–21).[18] Similarly, the purpose of the creation is to guide humans and to test their faith and conduct. Creation points to Allah and shows people God's will (6:165; 11:7; 14:10; 43:10 and 27; and 76:2).[19]

Another purpose of creation is to serve Allah and to be submissive to God's will. One way that God's creatures do this is to fight evil in the world. The insistence that this serving Allah is to help prepare for Judgment Day demonstrates once again the eschatological perspective of Islam (2:21; 10:6 and 31; 13:2; 51:56; and 67:5).[20]

Finally, the world is said to be created for the use of humans. Sometimes this suggests simply and solely that creation is intended to meet human needs. At other times, however, this view undergirds the emphasis on Allah's role as judge, for Allah gives humans the creation for their use in order to test how well they use it (2:22; 13:17; 14:32–33; 16:5–16 and 80–81; 17:70; 21:31–32; 23:18–22; 43:10–12; 55:1–78; and 78:6–16). To a lesser degree, the world is said to be created for the use of all creatures, not only humans (25:47–49; 55:10; and 80:24–32).[21]

The view of the world being created for human purposes, however, predominates in the early Islamic tradition.

The Theological Implications of Creation

What are the theological implications of the understanding of creation in the early Islamic tradition? A frequently recurring phrase in the Qur'an refers to the sovereign Allah, "to Whom do belong all things in the heavens and on earth!" (14:2; similarly 2:116; 11:123; 16:52; 21:19; 23:84–89; and 31:26). Since everything belongs to Allah, the creation may be given to whomever God sees fit. Consistent with the eschatological character of Islam, Allah gives the gifts of the creation according to God's role as judge, rewarding believers and punishing the ungrateful or unbelievers (7:128; 10:4; and 53:31; and *Mishkat*, 502). Not only does all belong to God, but it is also dependent on Allah. The relationship between Creator and creatures is asymmetrical, then, for a gap remains between them (11:6 and 41:37).[22]

The creation may be given to whomever God has seen fit, but Allah has seen fit to give it to humans. That is, humans are seen as vicegerents for Allah on the earth *(khulafa' Allah)*. In Sura 22, immediately after affirming that everything belongs to Allah, the question is asked, "Seest thou not that God has made subject to you . . . all that is on the earth?" (22:65). Yes, Allah it is "Who has made you (His) vicegerents of the earth" (6:165).[23] The Qur'an, however, does not speak about the vicegerency of humans without a certain ambivalence, for as Allah says, "We did indeed offer the Trust to the Heavens and the Earth and the Mountains; but they refused to undertake it . . .; But man undertook it;—he was indeed unjust and foolish" (33:72; cf. 2:30 and 7:10).

Even though humans have on occasion proved to be unjust and foolish, human life seems to be valued relatively more highly than other forms of life (17:70).[24] This is clearly the case in regard to other animals. While the Qur'an and the *hadith* literature do express an appreciation for the value of nonhuman animal life, as I shall point out more fully below, they subordinate it to human life. For example, passages in the *hadith* literature allow the killing of animals which are not desirable to humans—such as rats, crows, and certain kinds of dogs and snakes (Cf. al-Bukhari, vol. 4, bk. 54, ch. 14–16, 334–36 and 339–40; and *Mishkat*, 763, 876, 879, and 881–82). Likewise human life is sometimes, but not always, valued more highly even than spiritual beings, such as angels and the jinn. This is symbolized especially in those passages where the angels are commanded to prostrate themselves before Adam (2:34; 7:11; 15:30; and 38:72–74).[25]

Perhaps most significantly the early Islamic understanding of creation implies that creation's primary function is religious, or devotional—that is, Allah's creation is good and appropriately prompts a response of praise (3:191; 7:54; 23:14; 32:7; 40:64; and 67:3).[26] This "devotional" motif is interwoven even in those passages that affirm the role of humans as Allah's vicegerents. Humans are expected to respond to the honor Allah pays them with worship and gratitude. The gift of vicegerency is a test—a test to see if humans obey the will of Allah in an attitude of humility, repentance, loyalty, and faith (2:21; 10:31; 39:6–7; 51:56; and 56:57).[27] The prime response expected, however, is gratitude. As indicated above, one purpose of creation is to provide signs of Allah's sovereignty and beneficence. These signs are intended less to provide an intellectual proof for the creative activity of God than to evoke gratitude in humans as they observe the wonders of creation (7:10; 16:78; 22:36–37; 25:50; 36:73; and 45:12).[28] The primary descriptive word for those who ignore or turn their backs on Allah is "ingrate" rather than "unbeliever."[29]

One should note that this devotional motif applies to the non-human creation too. "Whatever is in the heavens and on earth,— let it declare the Praises and Glory of God: for He is the Exalted in Might, the Wise" (57:1; the same or similar verse can also be found in the first verses of Suras 59, 61, 62, and 64). From the thunder to the angels, from the birds to the hills—all creation is to join in the praise of Allah (13:13; 16:49; 21:79; 24:41; 38:18–20; 55:6 and 29; and 59:24).[30] Perhaps this may be viewed merely as devotional poetry. Yet it suggests that all of creation should be respected by humans, because it too is to join them in worshiping Allah. The early Islamic tradition, then, values all life, not just human life. Animals are to be treated with compassion and concern, partly because they are God's creation, partly because they have a role to play in the devotional function of all creation (6:38; 11:64–65; 26:155–58; 54:27–30; and 91:11–14).[31] Perhaps, to complete the picture, it should be added that the support for humane treatment of animals is based sometimes on their usefulness to humans—or their usefulness to Islam, as shown by the value placed on horses because of their role in jihad.[32] In general, however, a creation that bears signs of God's power and grace is a creation that is to be treated with care and respect, lest one show disrespect and ingratitude to its Creator.

Ecological Consequences in Principle

What, then, may be the "ecological fallout" of the views of creation in the Qur'an and *hadith* literature? The answer is ambiguous. On the one hand, some material may support an anthropocentric approach to the environment, exploiting it for human purposes. As I have shown above, the early Islamic tradition clearly sees the heavens and the earth as created to serve human purposes. From this point of view the nonhuman creation will likely be appreciated, not for its intrinsic value, but only for its instrumental value for humans. Moreover, Allah is seen as elevating humans to vicegerency over the earth, thus possibly giving humans the right to control the environment according to their will. Allah delegates sovereignty over creation to humans, who now have functional authority over the rest of creation as *khulafa' Allah*. This strand of the early Islamic tradition does not bode well for the environment.

On the other hand, I believe that the Islamic emphasis on divine sovereignty outweighs this possibly detrimental understanding of human vicegerency. Allah bestows authority over creation on humans, not as an absolute right to do as they please, but as a test— a test of their obedience, loyalty, and gratitude to God. Abusing the earth violates the will of Allah; caring for it fulfills God's will. From this perspective human vicegerency means exercising responsible care for the environment, not violating or exploiting it.[33]

Respect for the environment follows from two other aspects of Islamic thought discussed above: the belief that the creation provides humans with "signs" of the sovereignty and grace of Allah and the belief that the nonhuman creation is ordained to praise Allah along with humans. These both support a point of view which ascribes intrinsic value to creation, deserving respect in its own right because it either points to or praises Allah. The spotlight of creation is not focused on humans alone, but humans and the rest of creation all have their mutually important, God-given roles to play in the ongoing drama of Allah's continuing creation.

I referred above to the "eschatological flavor" of Islam. According to conventional wisdom, eschatologically oriented religion neglects the environment, for why should one worry about preserving a few birds or trees or about cleaning up streams or the air when the main concern of religious life is to prepare for the next life, in the Paradise to come? Such an attitude does not follow from what is said in the early Islamic tradition about Allah as final judge, for how a person treats God's creation is part of the basis upon which one is judged by Allah. Since humans are given vicegerency as a test, how well they care for the earth that is their responsibility determines how well they measure up to Allah's expectations. Here the eschatological character of Islam reinforces present responsibility for the sake of future reward. Humans are to care for the earth now as part of the unfolding of Allah's continuing creation so that they may participate in the final Paradise.

In short, the potential effect of early Islam on the environment depends in large measure on how the *khilafat Allah* given to humans is interpreted. If it is interpreted in an anthropocentric way, seeing the purpose of creation as serving humans, then the result may be exploitation of the earth. If, however, the vicegerency of humans is seen as ultimately subordinate to divine sovereignty and will, then human authority over the creation becomes responsibility to care gratefully for the environment that belongs to God and serves Allah's will.

Do the Qur'an and the *hadith* literature support exploitation of the earth and its resources? My analysis has shown, I believe, a basic ambiguity in this literature. Surely there are passages that emphasize human authority over the earth, passages that might be interpreted to support exploitation of the earth for human purposes. Ultimately, however, I think that Islam's emphasis on the sovereignty of Allah countermands an interpretation of the tradition that allows for environmental exploitation. The creation serves human purposes, yes, but within limits. Those limits are defined by God's will for humans and for creation and by the positive value Allah has placed on creation, especially by virtue of its function of providing signs of the power and mercy of God. In early Islam, then, affirmation of divine majesty offers a basis for advocating responsible caring for the earth.

Ecological Consequences in Practice

Determining the actual "ecological fallout" of Islamic beliefs about God as Creator is a difficult matter. Motivations for human behavior are many and varied, and assessing the impact of religious beliefs on behavior, especially in the past, can be highly speculative. Federico Peirone claims, nonetheless, that the ecology of the Mediterranean world was profoundly influenced by the spread of Islam around its shores. Prohibitions of alcoholic drinks and pork led to the decrease of vineyards in a number of Mediterranean countries and to the substitution for pigs of sheep and goats, whose grazing patterns have had negative consequences for the region's once vast greenbelts. The flourishing of horses, camels, and peanuts in the region he also traces to Muslim influence.[34] In some ways, then, the consequences of Muslim beliefs for the environment can be detected.

The consequences of such beliefs in the present may be even more difficult to detect, for influences on decisions and actions in contemporary Muslim countries may be quite complex. If the early Islamic tradition, still authoritative for Muslims today, encourages responsible care for the environment, then why do these countries often have such severe problems with pollution and depletion of natural resources?

One reason may be that in Muslim countries it is primarily poor people who cause pollution and other damage to the environment. Therefore in such countries concern for economic development or even mere survival has greater priority for public policy than do environmental concerns. Worrying about ecological sustainability may seem to be a luxury when the demands for simple justice and fair distribution of economic resources to all people, especially those living in abject poverty, command the attention of a country's leaders. Moreover, in situations where primarily the poor cause and suffer from environmental degradation, people of wealth, power, or prestige are seldom affected by ecological problems and have little incentive to support efforts to protect the environment.[35]

Another reason may be the impact of modern, Western science and technology. As early Muslim thinkers assimilated Greek philosophy and science and as in turn medieval Christians borrowed from the flourishing of Islamic science, so in more recent times the Muslim world has been profoundly influenced by the success of European and American science and technology. Because of this influence and a concomitant secularization of their societies, the leaders of Muslim nations may be alienated from their religious roots, and the support in the Islamic worldview for caring for the natural environment may have little power to overcome the lure of Western technology and its frequently negative effects on the environment.[36]

Signs that the Islamic tradition does support caring for the environment can be found. Setting aside land to protect it from development, sometimes for use as wildlife preserves, is an ancient Muslim custom still observed in many Muslim countries. Government agencies, such as Saudi Arabia's Meteorology and Environmental Protection Administration, have been established to encourage adherence to Islamic principles of environmental protection. While dealing with the theme, "The Environmental Aspects of Development," the Arab Ministerial Conference in 1986 did consider the Islamic faith and its values in relation to ecological concerns.[37]

Islam's monotheistic creation theology does not necessarily support exploitation of the earth. Resources for encouraging responsible care of the environment clearly exist in the Islamic tradition. The challenge for those within the Muslim world who are sensitive to ecological issues is to highlight these aspects of their faith tradition and to call on Muslims to live up to the responsibility for caring for Allah's creation that their faith entrusts to them.

Notes

1. Perhaps the classic locus for this argument is stated in the article by Lynn White, Jr., "The Historical Roots of Our Ecologic Crisis," *Science*, no. 155 (1967), 1203–7.

2. By "early Islam" or "the early Islamic religious tradition" I intend to refer to Islamic beliefs and practices as expressed in the Qur'an and the *hadith* literature. I attempt to base my claims first of all on passages from the Qur'an, secondly on material from the *hadith* literature, and thirdly on secondary sources; my references are listed in that order. I recognize that *hadith* literature is both vast and often of questionable authenticity; I rely primarily on the *hadith* literature collected by al-Bukhari, considered the most reliable collector of *hadith*, and add references to other *hadith* literature only as supplementary information.

3. I am not interested in arguing that Islam is somehow dependent on either the Jewish or Christian religious traditions; for the purposes of this paper I am quite content to accept the traditional Muslim belief in the relative independence of Muhammad and the revelations contained in the Qur'an. Even in the Qur'an, however, a connection with the biblical traditions is clearly acknowledged, for Muhammad is seen as being the culmination of a line of prophets stretching from Abraham through Moses and Jesus. See, for example, Suras 6:83–90; 33:7–8, and 51:24–46. My references to the Qur'an are drawn from *The Holy Qur'an*, trans. Abdullah Yusuf 'Ali, 3d ed. (Washington, D.C.: The Islamic Center, 1938) and will be cited in the text by Sura numbers and verses.

4. A somewhat expanded description of the references to creation in the Qur'an and *hadith* literature can be found in my article "Divine Majesty, Human Vicegerency, and the Fate of the Earth in Early Islam," *Hamdard Islamicus* 13 (Spring 1990): 47–57, which is also included in *Essays on Islam*, ed. Hakim Mohammed Said (Karachi: Hamdard Foundation Pakistan, 1992), 1:1–15.

5. See also Muhammad ibn Isma'il al-Bukhari, *The Translation of the Meanings of Sahih*

Al-Bukhari, trans. Muhammad Muhsin Khan, 3d ed. (Chicago: Kazi Publications, 1979), vol. 6, bk. 60, ch. 255, 322–23; hereafter cited as al-Bukhari.

6. See R. Arnaldez, *"Khalk," The Encyclopedia of Islam,* ed. E. van Donzel et al., new ed. (Leiden: Brill, 1978), 4:984. A story collected in the *hadith* literature that seems to give a seven day chronology for creation may be making a similar point, for rather than resting on the seventh day Allah creates Adam; cf. Wali ad-Din Muhammad ibn 'Abdallah al-Khatib at-Tibrizi, *Mishkat al-Masabih,* trans. James Robson (Lahore: Sh. Muhammad Ashraf, 1965), 1227–28; hereafter cited as *Mishkat.*

7. But cf. al-Bukhari, vol. 4, bk. 55, ch. 1, 345.

8. See *Mishkat,* 506; Richard Bell, *Introduction to the Qur'an,* rev. W. Montgomery Watt (Edinburgh: Edinburgh University Press, 1970), 149–50, hereafter cited as Bell; Helmut Gaetje, *The Qur'an and Its Exegesis: Selected Texts with Classical and Modern Muslim Interpretations,* trans. Alford T. Welch (Berkeley: University of California Press, 1976), 149 (quoting the Muslim Quranic commentator, al-Baidawi); and Fazlur Rahman, *Major Themes of the Qur'an* (Minneapolis: Bibliotheca Islamica, 1980), 7 and 67.

9. Note, however, that the pre-Islamic Arabs, even though polytheistic, usually considered only the highest God, Allah, as the Creator; cf. Toshihiko Izutsu, *God and Man in the Koran: Semantics of the Koranic Weltanschauung* (Tokyo: Keio Institute of Cultural and Linguistic Studies, 1964), 121.

10. Muhammad's customary morning and evening prayers attributed dominion over the world to Allah; cf. *Mishkat,* 506. See also J. M. S. Baljon, "The 'Amr of God' in the Koran," *Acta Orientalia* (Copenhagen) 23 (1958):8–11; Gaetje, *Exegesis,* 149–51; Siegfried Raeder, "Sure 55: Schoepfung und Gericht: Versuch einer religionswissenschaftlichen Deutung und theologischen Kritik," *Prophetie in Bibel und Koran,* ed. Willi Hoepfner (Wiesbaden: Verlag der Evangelischen Mission in Oberaegypten, 1974), 46–47; and Rahman, *Major Themes,* 1, 6 and 65–66. Sheila McDonough in "The Qur'an and Patriarchal Religion" suggests that the portrayal of Allah as male in the Qur'an stems in part from emphasizing God's omnipotence, for in Arabian culture of that time power was seen as a masculine characteristic; see *Sciences Religieuses/Studies in Religion* 6 (1976–77): 538.

11. See also *Mishkat,* 1219; Toshihiko Izutsu, *Ethico-Religious Concepts in the Qur'an* (Montreal: McGill University Press, 1966), 120; Izutsu, *God and Man,* 22–23 and 120; and Rahman, *Major Themes,* 1–3, 6, and 65–66.

12. See Izutsu, *God and Man,* 129; Thomas J. O'Shaughnessy, "God's Purpose in Creating According to the Qur'an," *Journal of Semitic Studies* 20 (1975): 203–4; and Raeder, "Sure 55," 45–46.

13. See Arnaldez, *"Khalk,"* 985; Baljon, "'Amr of God,'" 9; Fritz Meier, "The Ultimate Origin and the Hereafter in Islam," in *Islam and Its Cultural Divergence: Studies in Honor of Gustave E. von Grunebaum,* ed. Girdhari L. Tikku (Urbana: University of Illinois Press, 1971), 103; and Annemarie Schimmel, "Creation and Judgment in the Koran and in Mystico-Poetical Interpretation," in *We Believe in One God: The Experience of God in Christianity and Islam,* ed. Annemarie Schimmel and Abdoldjavad Falaturi (New York: Seabury Press, 1979), 150–51.

14. See Izutsu, *God and Man,* 129; O'Shaughnessy, "God's Purpose," 200; Rahman, *Major Themes,* 1; and Schimmel, "Creation and Judgment," 150.

15. See Arnaldez, *"Khalk,"* 983; and O'Shaughnessy, "God's Purpose," 197.

16. See *Mishkat,* 1168–69; Theodor Lohmann, "Schoepfung und Auferweckung des Menschen im Koran," in *Beitraege zu Geschichte, Kultur und Religion des alten Orients,* ed. Manfred Lurker (Baden-Baden: Koerner, 1971), 254–55 and 259; and Raeder, "Sure 55," 43–44.

17. See Arnaldez, *"Khalk,"* 982, and Muin-ud-Din Ahmad Khan, "Reflection on the Quranic Concept of the Creation of Universe and Mankind," *International Islamic Conference*

(February 1968), vol. 1: *English Papers* (Islamabad, Pakistan: Islamic Research Institute, 1970), 32. Similarly, when the Qur'an states that Allah created the world "in truth," it seems to be affirming that the world was created purposefully; cf. Sura 6:73; 15:85; 30:8; 45:22; and 64:3. For a general discussion of the purpose of creation see O'Shaughnessy, "God's Purpose," and Rahman, *Major Themes*, 1–3, 7–14, 32–34, and 66–79. By "the creation" here I mean *what* God created, *not* God's *act* of creating.

18. See al-Bukhari, vol. 4, bk. 54, ch. 4, 284–85; and vol. 6, bk. 60, ch. 69, 75; Bell, *Introduction*, 122 and 148–49; Gaetje, *Exegesis*, 152–53 (quoting Baidawi on Sura 2:164); and Izutsu, *God and Man*, 133–35 and 231.

19. See al-Bukhari, vol. 4, bk. 54, ch. 3, 282; and Kahn, "Reflection," 33–34.

20. See al-Bukhari, vol. 4, bk. 55, ch. 1, 347; Abdoldjavad Falaturi, "How Can a Muslim Experience God, Given Islam's Radical Monotheism?" in Schimmel and Falaturi, *One God*, 81–83 and 87; and Schimmel, "Creation and Judgment," 155.

21. Cf. *Mishkat*, 634, where a *hadith* transmitted by Abu Dawud states that someone will be punished who unjustifiably cuts down a tree giving shade, not only to humans, but also to animals.

22. See Izutsu, *Ethico-Religious Concepts*, 149 and 151; Izutsu, *God and Man*, 75 and 122; Raeder, "Sure 55," 46–47; and Rahman, *Major Themes*, 2, 10, 67, and 167.

23. See Sura 2:30; 7:10 and 69; 10:14 and 73; 27:62; 35:39; and 36:71–72; *Mishkat*, 1067; Arnaldez, "*Khalk*," 982; and Khan, "Reflection," 33–35.

24. See *Mishkat*, 455; and Rahman, *Major Themes*, 46. Sometimes, however, to encourage humility the creation of humans is described as being less important than the creation of the heavens and the earth; cf. Sura 40:57 and 95:4–6.

25. See *Mishkat*, 1227; Gaetje, *Exegesis*, 168 (quoting Zamakhshari on Sura 38); Meier, "Ultimate Origin," 99; and Rahman, *Major Themes*, 17–18 and 167.

26. See *Mishkat*, 165–66 and 1067; and Gaetje, *Exegesis*, 150 and 152 (Baidawi on Sura 7:54, and Zamakhshari on Sura 32:7).

27. See al-Bukhari, vol. 4, bk. 55, ch. 1, 347; Bell, *Introduction*, 122; Khan, "Reflection," 33–35; Raeder, "Sure 55," 47 and 50; and Rahman, *Major Themes*, 12, 14, 18–19, and 79.

28. See *Mishkat*, 33; and Izutsu, *God and Man*, 231.

29. See Sura 16:83; Bell, *Introduction*, 150; Izutsu, *Ethico-Religious Concepts*, 120 and 200; and Izutsu, *God and Man*, 22–23 and 77.

30. See al-Bukhari, vol. 4, bk. 52, ch. 153, 162; vol. 4, bk. 54. ch. 14, 332; and vol. 6, bk. 60, ch. 247, 309–10; *Mishkat*, 879–81; Falaturi, "Experience God?" 86; Schimmel, "Creation and Judgment," 155; and Federico Peirone, "Islam and Ecology in the Mediterranean Muslim *Kulturkreise*," *Hamdard Islamicus* 5, no. 2 (Summer 1982): 27.

31. See al-Bukhari, vol. 4, bk. 52, ch. 86, 103; and ch. 154, 163; vol. 4, bk. 54, ch. 6, 297–99; and ch. 15–16, 337–39; *Mishkat*, 717, 872, 874, and 882; Maulana Muhammad Ali, ed., *A Manual of Hadith* (Lahore: Ahmadiyya Anjuman Ishaat-I-Islam, n.d.), 344–45 and 348–49; Alfred Guillaume, *The Traditions of Islam: An Introduction to the Study of the Hadith Literature* (Oxford: Clarendon Press, 1924), 106–7; and Peirone, "Islam and Ecology," 24–27.

32. See Sura 6:38; al-Bukhari, vol. 4, bk. 52, ch. 43, 71; and ch. 48, 75–76; vol. 4, bk. 54, ch. 13, 330–31; and ch. 14–15, 335–36; and vol. 6, bk. 60, ch. 350, 458–59 (on Sura 99:7–8); and *Mishkat*, 763, 876, and 881–82.

33. See Horst Buerkle, "Secularization—A Theme in Christianity's Dialogue with Non-Christian Religions," in Schimmel and Falaturi, *One God*, 129; and Iqtidar H. Zaidi, "On the Ethics of Man's Interaction with the Environment: An Islamic Approach," in *Religion and Environmental Crisis*, ed. Eugene C. Hargrove (Athens: University of Georgia Press, 1986), 113–16.

34. Peirone, "Islam and Ecology," 19–31.

35. O. A. El Kholy, "Science, Technology and the Future: An Arab Perspective," in *Faith and Science in an Unjust World,* ed. Roger L. Shinn (Philadelphia: Fortress Press, 1980), 1:136–38; and Lloyd Timberlake, "The Emergence of Environmental Awareness in the West," in *The Touch of Midas: Science, Values and Environment in Islam and the West,* ed. Sardar Ziauddin (Manchester: Manchester University Press, 1984), 127–30.

36. Timberlake, "Emergence," 130–33; and S. Parvez Manzoor, "Environment and Values: the Islamic Perspective," *Touch of Midas,* 150–51 and 154.

37. Mawil Y. Izzi Deen (Samarrai), "Islamic Environmental Ethics, Law and Society," in *Ethics of Environment and Development: Global Challenge, International Response,* ed. J. Ronald Engel and Joan Gibb Engel (Tucson: University of Arizona Press, 1990), 196–97. See the summary of Islamic environmental principles written to guide the formation and operation of Saudi Arabia's Meteorology and Environmental Protection Administration: Abou Bakr Ahmed Ba Kader et al., *Islamic Principles for the Conservation of the Natural Environment* (Gland, Switzerland: International Union for Conservation of Nature and Natural Resources, 1983). In addition there is the important work of Seyyed Hussein Nasr as seen in his book *Man and Nature: The Spiritual Crisis and Modern Man,* first published in 1968 and reissued in paperback in 1976 and 1988 by Unwin in London.

A Baha'i Perspective on an Ecologically Sustainable Society

Robert A. White

THIS essay takes a broad macroevolutionary approach to our changing relationship to nature in light of the teachings of the Baha'i Faith. It suggests that humanity is in a process of evolving consciousness that is leading to the development of a new planetary culture which will be based on a mature cooperative relationship between humanity and the ecosphere that gave it birth. This paper first explores the basic attitudes to nature that are contained within the Baha'i writings and then examines how the emergence of an ecologically sustainable social order is linked to some of the principles of the Baha'i Faith. Implicit throughout is the Baha'i view of the balance and cohesion of material and spiritual realities. While references are made to Baha'i texts, the interpretation presented is a personal one.

Relationship with Nature in Revision

The cumulative breakdown of the relationship between the human species and the ecosphere has reached a point at which mere technical and social adjustments to prevailing models of development are inadequate to ensure planetary sustainability. The call for a "radically new metaphysic"[1] that recognizes the reciprocal relationship between humanity and nature has spawned various schools of thought such as "deep ecology," "ecophilosophy," and "ecofeminism." With an even more comprehensive vision Thomas Berry eloquently describes the essential dimensions of an "Ecological Age" into which we are now moving.[2] All of these schools of thought call, in one form or another, for a transformation of consciousness away from seeing the earth as a collection of resources to be exploited and consumed to one of humanity living as an integral, cocreative part of the ecosphere. This fundamental change involves an appreciation of the spiritual dimension as a necessary element of our relationship to nature both individually and collectively.

It is within this context that the teachings of the Baha'i Faith make a significant contribution. In their emphasis on unity and evolutionary thinking, the Baha'i teachings offer a view of nature that embraces both animistic wisdom and contemporary ecological understanding. At the same time, these teachings affirm divine transcendence and the essential unity of religious expression throughout history. Furthermore, they present a challenging interpretation of what religion is and its role in transforming the current world order. In addition, many of the tenets and principles for an alternative society based on ecological wisdom are found within the writings and institutions of the Baha'i Faith.

Relationship with Nature: A Baha'i Perspective

In an examination of the principles of the Baha'i Faith as they apply to agriculture, Paul Hanley articulates a threefold relationship between humanity and nature involving principles of unity, detachment, and humility.[3] These same principles will be explored in depth here.

Unity with Nature

'Abdu'l-Bahá asserts that "all parts of the creational world are of one whole. . . . All the parts are subordinate and obedient to the whole. The contingent beings are the branches of the tree of life while the Messenger of God is the root of that tree."[4] He illustrates this essential unity in the following analogy:

> Liken the world of existence to the temple of man. All the limbs and organs of the human body assist one another; therefore life continues. . . . Likewise, among the parts of existence there is a wonderful connection and interchange of forces, which is the cause of the life of the world and the continuation of these countless phenomena. . . . From this illustration one can see the base of life is this mutual aid and helpfulness.[5]

According to 'Abdu'l-Bahá, the cooperative interrelations of creation are a manifestation of love, which is "the secret of God's holy Dispensation." Through God's love the world of being receives life:

> Love is the cause of God's revelation unto man, the vital bond inherent, in accordance with the divine creation, in the realities of things . . . Love is the most great law that ruleth this mighty and heavenly cycle, the unique power that bindeth together the divers elements of this material world, the supreme magnetic force that directeth the movements of the spheres in the celestial realms.[6]

Further, the mineral, plant, and animal are seen to possess various grades and stations of spirit. 'Abdu'l-Bahá wrote in 1921:

> it is indubitable that minerals are endowed with a spirit and life according to the requirements of that stage. . . .
> In the vegetable world, too, there is the power of growth, and that power of growth is the spirit. In the animal world there is the sense of feeling, but in the human world there is an all-embracing power . . . the reasoning power of the mind. . . .
> In like manner the mind proveth the existence of an unseen Reality that embraceth all beings, and that existeth and revealeth itself in all stages.[7]

There is a cohesiveness within life's ever-increasing differentiation—an underlying spirit that animates all of existence. The prevailing view of nature as "environment" made up of material components of air, water, soil, and organisms is therefore inadequate. The very word "environment" implies that which is external and peripheral to what is assumed to be the central object of concern, human beings. This human self-preoccupation ignores the reality that life and spirit are properties of the whole and its reciprocal interactions.

Spiritual Detachment from Nature

Humanity is part of the whole of creation which in turn reflects, in its harmony and unity, a divine and unseen reality. At the same time, paradoxically, human beings occupy a unique station that can be consciously realized only through detachment from nature. 'Abdu'l-Bahá states that the human being "is in the highest degree of materiality, and at the beginning of spirituality."[8]

Creation is seen as a progression of increasingly complex orders from the mineral kingdom to vegetable and animal life to human beings. Humanity, however, has the capability and the power of spiritual advancement, our very purpose being to advance toward God. As stated by 'Abdu'l-Bahá:

> God has created all earthly things under a law of progression in material degrees, but He has created man and endowed him with powers of advancement toward spiritual and transcendental kingdoms.[9]

All other created things are "captives of nature and the sense world," but human beings, created in the "image of God," occupy a unique station in creation. We have evolved through all the physical kingdoms and contain all of their capacities plus the distinguishing capacity for rational and self-reflective thought. The development of this unique capacity of the mind that allows us to mediate between the material and spiritual dimensions has required us to separate ourselves from nature both externally and internally. Through this separation humanity has gained the capacity to know nature from the outside and to unravel its secrets. In an internal sense we have partially removed ourselves from the physical and instinctual responses that guide all other life forms and have developed conscious faculties of judgment and volition.

The freedom these capacities give us involves a commensurate responsibility to recognize the "unseen Reality that embraceth all beings" (*BR*, 222). Our spiritual evolution depends on the degree of our attunement to that greater reality, which is described by Bahá'u'lláh and all the great prophets as limitless and eternal. Thus, to truly develop a conscious spirituality and to awaken to our full potential we are called to sever our immediate identification with the physical dimension of nature. 'Abdu'l-Bahá discusses this concept:

> And among the teachings of His Holiness Bahá'u'lláh is man's freedom, that through the ideal Power he should be free and emancipated from the captivity of the world of nature. . . . Until man is born again from the world of nature, that is to say, becomes detached from the world of nature, he is essentially an animal, and it is the teachings of God which converts this animal into a human soul. (*BWF*, 288–90)

What is problematic is an absorption in the material, as an end in itself. Detachment from the physical world is a means of gaining conscious access to the spiritual realities that lie behind and beyond the physical. Paradoxically, this detachment allows us to see that the physical world perfectly and fully reflects the spiritual world. This is demonstrated, as John Hatcher points out, in our growing awareness of ecology.[10] As we begin to understand the ecological principle that everything is connected to everything else in the physical world, we are learning the essential truth of the spiritual law of unity that pervades and animates all of creation.

The paradox between our unity and our detachment can be seen on deeper reflection as representing the multidimensionality of our humanness. The recognition of our unity with the earth, which in a very real sense gestated us, reflects both animistic wisdom and contemporary ecological understanding. At the same time, as earlier religious revelations emphasized, we must reach beyond the material world to discover our spiritual potential and to fulfill our destiny as conscious beings. That potential and destiny, which has been reflected to us by a progression of divine messengers, is an unfolding one in an ongoing process of creation. Faith in and vision of our perfectibility gives us the strength to progress toward fulfillment of all our potential and to participate in spiritualizing our social existence.

While the Baha'i Faith is not the first belief system to recognize this tension between the material and spiritual dimensions, it brings a fuller appreciation of the balance in this relationship. Matthew Fox seeks just such a balance in his call for "panentheism."[11] Like pantheism, panentheism sees the spirit of God as present in all things; at the same time, God is an independent Being above and beyond all things. Similarly, Bahá'u'lláh has written:

> The whole universe reflecteth His glory, while he is Himself independent of, and transcendeth His creatures. This is the true meaning of Divine unity. He Who is the Eternal Truth is the one Power Who exerciseth undisputed sovereignty over the world of being, Whose image is reflected in the mirror of the entire creation."[12]

Humility

In this delicate balance between unity and detachment, we are called on to honor creation, to recognize its sacredness, and to humble ourselves before it. In the miracle of life's evolution, God has acted through nature in an "emergent" way. Creation is intrinsically endowed with meaning and purpose, and reflects the unity, beauty, and ultimate mystery of God. The earth itself reveals the attributes of God as Bahá'u'lláh affirms:

> How all-encompassing are the wonders of His boundless grace! Behold how they have pervaded the whole of creation. Such is their

virtue that not a single atom in the entire universe can be found which doth not declare the evidences of His might, which doth not glorify His holy Name, or is not expressive of the effulgent light of His unity. So perfect and comprehensive is His creation that no mind nor heart, however keen or pure, can ever grasp the nature of the most insignificant of His creatures; much less fathom the mystery of Him Who is the Day Star of Truth, Who is the invisible and unknowable Essence. (*G*, 62)

'Abdu'l-Bahá describes creation as one of the "two Books" of God. "The Book of Creation is in accord with the written Book"— the sacred revelations of all the prophets of God. Like the written book, "The Book of Creation is the command of God and the repository of divine mysteries."[13]

The spirituality of the world's aboriginal cultures is based on understanding the primary "scripture" of the "Book of Creation"; and in the revealed religions, symbols of nature such as trees, water, and mountains also carry spiritual meaning. Both by direct contact and through symbolic reflection, the human soul is nourished by connection with the beauty, mystery, and grandeur of nature.

An attitude of awe and gratitude toward the earth is part of attaining spiritual humility. Humility means literally of the ground or humus. Bahá'u'lláh describes this relationship:

Every man of discernment, while walking upon the earth, feeleth indeed abashed, inasmuch as he is fully aware that the thing which is the source of his prosperity, his wealth, his might, his exaltation, his advancement and power is, as ordained by God, the very earth which is trodden beneath the feet of all men. There can be no doubt that whoever is cognizant of this truth, is cleansed and sanctified from all pride, arrogance, and vainglory.[14]

A New Vision of Wholeness in Our Relationship to Nature

Developing new attitudes of respect for and cooperation with nature requires, first of all, a vision of wholeness in our relationship to nature. This requires a perspective of human evolution and human purpose that unifies material and spiritual realities. The focus on transcending nature, which has characterized Western civilization in particular, is reflected in the current species' self-centeredness of the human race. The divorce of human destiny from the reality of physical life on earth now requires a reconciliation. However, this cannot be achieved through the replacement of our anthropocentrism by a biocentrism. Rather, our separation and detachment from nature and our unity with it must be understood as a creative dialectic in the development of human consciousness.

The process of becoming conscious beings has required us to break away from our unconscious roots in nature and to identify with a vision of our potential that transcends the physical. This separation has left us with no secure grounding for who we are and no clear vision of our wholeness. We retain only a dim memory of our unconscious unity with nature and a vague hope for the restoration of peace and wholeness in an abstract heaven or a future kingdom of God. The negative self-concept we hold as fallen creatures itself breeds guilt, despair, and abasement of both ourselves and creation. However, the Baha'i writings make it clear that we came into being in a perfect creation and that our station in it is a noble one. We are the "fruit of creation," conscious beings given the responsibility of fulfilling creation by reflecting its perfections. 'Abdu'l-Bahá addresses this situation as follows:

> One of the things which has appeared in the world of existence, and which is one of the requirements of Nature, is human life. Considered from this point of view man is the branch; nature is the root. Then can the will and the intelligence, and the perfections which exist in the branch be absent in the root? (SAQ, 4)

He further states that humanity "in the body of the world is like the brain and mind in man . . . man is the greatest member of this world, and if the body was without this chief member, surely it would be imperfect. We consider man as the greatest member, because, among the creatures, he is the sum of all existing perfections" (SAQ, 178). Bahá'u'lláh addresses the same theme:

> To a supreme degree is this true of man, who, among all created things, hath been invested with the robe of such gifts, and hath been singled out for the glory of such distinction. For in him are potentially revealed all the attributes and names of God to a degree that no other created being hath excelled or surpassed. (G, 177)

We are, in other words, nature becoming conscious of itself, but the gift of that consciousness lifts us into another dimension. Nature is perfect in itself because it is governed by laws and rules ordained by God. This perfection is reflected in all the metaphors of nature used in the writings of Bahá'u'lláh and earlier prophets. The perfection of human beings, however, are unrealized. We must choose to realize them through the development of our latent spiritual capacities. These capacities to reveal the "attributes and names of God" are always evolving and are reflected to us by a series of divine messengers and their revelations. In the evolution of humanity toward conscious wholeness and completion, the messenger of God is the key to the union of material and spiritual realities. Thus the center of existence is neither humanity nor nature (neither anthropocentrism nor biocentrism). It is God through his manifestation that is the "root" of the "tree of life" (BWF, 364). In this era, Baha'is believe, the unification manifested by Bahá'u'lláh has released the potential for us to transform ourselves toward a more complete reflection of the perfections of God and the wholeness of creation.

In this light, the deepening crises of planetary destruction must be seen not as the inevitable failure of fallen humanity but as a crucial stage in the evolution of human consciousness toward greater wholeness. These crises impel us to reflect on the incompleteness of our current vision and respond with urgency to the forces of transformation. The second part of this paper will explore the social dimensions of this spiritual process of transformation.

Toward a Global Civilization: The Evolution of an Ecologically Sustainable Society

Understanding creation as sacred and whole and seeing the role of human beings to be conscious, compassionate, and creative participants in the evolution of life is the ultimate requirement for an ecologically sustainable society. However, developing this society will require not only a transformation in our individual attitudes and values but also a complete reordering of our social structures. Most of the socioeconomic institutions of modern industrial societies are based on the pursuit of material progress through separation from and conquest over nature. This separation denies a meaningful relatedness to the whole of creation and thereby denies sacredness to life. This loss of meaning and the ensuing emptiness fuel, in turn, the search for fulfillment through consumption, competition, and other addictive behaviors. The separation from nature underlying modernism corresponds to a division between mind and heart.

Incorporating a new vision of wholeness in our relationship to the earth requires a reincorporation of the spiritual dimension. Yet we cling, says Henryk Skolimowski, to our ideals of "secular salvation" because the successes of science and humanism seem too hard won to betray.[15] Despite this resistance, the prevailing worldview of secular materialism is being undermined both by the proliferation of its problems and contradictions and by the emergence of more inclusive cosmologies that provide new organizing principles. The unity of the material and spiritual dimensions is just such a principle that provides a foundation for a vision of humanity in relation to the whole of creation. Discoveries on the new frontiers of science point to this kind of integration and provide analogies such as in physics where light is understood as both a wave and a particle. The emerging world vision must similarly account for human beings as both biological and spiritual beings. Skolimowski asserts that humans

> are the custodians of the whole of evolution, and at the same time only the point on the arrow of evolution. . . . The sacredness of man is the uniqueness of his biological constitution which is endowed with such refined potentials that it can attain spirituality. (E-P, 74–75)

The Baha'i Faith is based on such an evolutionary perspective. As has been discussed earlier, it views our spiritual origin and destiny as part of the whole evolution of life on the planet. Spirit is expressed in all the stages and processes of existence and is realized consciously through the distinctive capacities of human awareness. The development of civilization itself is seen as a progressive organic process of maturation in which all the great religious revelations and scientific advancements are integral components. This dynamic and holistic perspective can help us pierce the socially constructed dichotomy of humanity versus planet and overcome the destructive divisions of the human race. In this larger evolutionary movement our current situation can be understood as a crucial stage in the birth of a new world order appropriate to humanity's spiritual and intellectual maturity. The teachings of the Baha'i Faith not only delineate the past and future dimensions of this process, they also offer values, principles, and new institutional forms that can guide us through this transition to maturity and the development of a global civilization.

Evolutionary Perspective

In the Baha'i writings the evolution of humanity is viewed as a purposeful organic process. Natural images, such as the earth developing in "the matrix of the universe" and the human species growing and developing in the "womb of the earth," are used to describe the nature of this process (*SAQ*, 182–83). The evolution of civilization is also viewed organically using the analogy of human developmental stages.[16]

Within this framework of maturation it is possible to sketch out humanity's changing relationship to the earth. In the earliest phases of the human journey, human dependence on the earth was embodied in myths and cultural forms. Symbols of the life-giving earth as "mother" signified the bonding typical of childhood. By association, the role of the "feminine" in its generative and nurturing aspects was generally accorded greater recognition.

In the emergence of the axial civilizations there was greater emphasis on rationality, independence, and order, representing a shift to the primacy of the "masculine." As a result, nature was gradually demythologized and spiritual and intellectual pursuits separated and abstracted from the instinctive primal energies of the body and the earth.

Western science developed in this context and took as its basic operating assumption the radical separation of subject and object, humanity and nature. The earth ceased to be a community to which humanity belonged and increasingly became a commodity for use and possession. As a result, our original dependence on the natural world has been replaced by alienation from nature and power over a meaningless material world.[17]

As destructive as this alienation has been in terms of the domination of nature, this mindset can be understood in the larger evolutionary context as a necessary phase of humanity's maturing consciousness. Just as in adolescence, when individuation requires the fragile ego to emerge and assert itself, so too the human race has had to break away from the primordial natural unity. However, to continue to assert an extreme degree of independence and "false sense of omnipotence" is no longer viable.[18]

Our evolutionary imperative is to leave this adolescent phase and progress to a mature understanding of our true interdependence with nature. This relationship will involve the integration of religious and scientific dimensions. Religion expresses the sense of being encompassed by something greater with which harmony is sought, while science expresses the urge to encompass and to know ("Science," 15–16). Maturity in our relationship to the earth implies a genuine humility about our dependent position in the ecosphere, while at the same time meeting human needs through applying skills and technologies based on ecological principles.

Science has itself become the means by which the truth of life's profound interdependence has come to light. The emergent paradigms in ecology, quantum physics, neurology, and psychology all point to the fundamental unity of life and to the interpenetration of matter and spirit. The fact that science is now confirming this dynamic interconnectedness of life does not by itself restore a subjective relatedness or sense of wholeness. Restoration of that wholeness on a conscious level is a process related to the root meaning of religion—to reconnect, or bind back. However, in this era of expanding human knowledge this healing reconnection requires a religious understanding that is complementary to the scientific principle that truth is relative and progressive.

In the nineteenth century, Bahá'u'lláh stated that the revelation of religious truth is an ongoing, open-ended process that has animated humanity's development toward greater unity and consciousness. Within this context, Bahá'u'lláh claimed that the role of his revelation was to initiate a process of conscious unification on a planetary scale—a process appropriate to humanity's maturation and growing technological capacities. The pivotal expression of this process is the recognition and acceptance of the unity of the human family. Planetary unity is the necessary and inevitable fruition of humanity's collective spiritual and material development—"the consummation of human evolution."[19]

The Unity Paradigm: Precondition for an Ecologically Sustainable Society

The oneness of humanity as enunciated by Bahá'u'lláh is not simply "an expression of vague and pious hope" or a renewal of traditional calls for solidarity. "It implies an organic change in the structure of present-day society, a change such as the world has not yet experienced" (WOB, 42–43). It calls for a reflection in the world of humanity of the fundamental law of oneness in the whole of creation. The Baha'i Faith, therefore, presents a unity paradigm which reflects an altered understanding of the relationship of parts to each other and to the whole. This new degree of integration is part of humanity's maturing consciousness following upon its entire developmental process and its increasing levels of interaction and interdependence.

This coming of age requires not just a perceptual shift; it requires institutional arrangements which acknowledge the primacy of the whole. Foremost among these is some form of world federal system guided by universally agreed upon values and laws, which can reflect the reality of humanity's oneness and its integral dependence on the encompassing ecosphere, which is itself a unified whole. Systems of thought and governance appropriate to humanity's adolescence must give way to new patterns and new institutions necessary to manage cooperatively an increasingly interdependent world.

Recognition of the consequences of disunity and the necessity of unity is a crucial component of this transition. The costs of nationalism, racism, and other forms of disunity can be tallied in the social and ecological effects of war, inequality, and grossly irresponsible military expenditure. The global military budget continues to run in excess of nine hundred billion dollars annually despite prospects of de-escalation due to the end of the cold war. Less than five percent of this amount (forty-five billion annually) would fund the most urgent global environmental requirements—preventing soil erosion, protecting and replanting forests, protecting the ozone layer, cleaning up hazardous wastes, developing renewable sources of energy, and stabilizing population.[20]

Current global issues, especially ecological concerns that transcend national boundaries, are, in effect, "forcing functions" requiring the community of nations to move beyond ad hoc and reactionary approaches to solving problems. The call for an integrated global ethic and policy of sustainable development raised in *Our Common Future,* and further amplified throughout the Earth Summit process, represents a growing acceptance of the need for unity in solving global problems. With this acceptance, there is a growing search for ways to bring about the changes in attitude and motivation required for unified global action. The creation of a Sustainable Development Commission by the United Nations, as part of the implementation of the Earth Summit's "Agenda 21," is one small but significant step in the recognition of the need for global goals and principles that anticipate and guide future development.[21]

While all such steps are important and necessary, political and social reorganization can only become effective to the degree that the consciousness of the oneness of humanity is the operating premise. This spiritual and organic truth, once accepted, can release the constructive energy that will be needed to make the far-reaching structural changes required for fostering sustainable patterns of development. The principle of unity must be the foundation for building and mandating institutions that can responsibly address environment and development issues on a global scale. It is for this reason that a major emphasis of the Baha'i International Community is on developing a consultative and institutional framework that demonstrates the viability of operating as a unified global community.

Globalism and Decentralism

The call by deep ecologists and other social theorists for decentralized, small-scale, community-based technologies and economies, at first glance, seems to represent the opposite extreme of the global thinking introduced above. "Ecological consciousness," it is reasoned, has mostly developed within a "minority tradition" that includes tribal cultures, utopian communities, and many monastic and religious communities.[22] There is a concern that a "global" society would become just a more effective superstate for the conquest of the ecosphere. What is needed, decentralists suggest, is to develop communities on an ecosystem-specific basis (bioregionalism) with people committed to "reinhabiting" and restoring that ecosystem and developing a renewed "sense of place."

There are several aspects of the Baha'i approach that relate to these decentralist concepts. First and foremost, the Baha'i concept of globalism is highly cognizant of the importance of traditional cultures and religions, while recognizing the need for global order and regulation. The Baha'i concept of globalism "repudiates excessive centralization on one hand, and disclaims all attempts at uniformity on the other. Its watchword is unity in diversity" (WOB, 42).

The present structure of the Baha'i International Community offers a pattern for constructing a worldwide society whose vision is world-embracing but whose members and activities are exceedingly diverse. It comprises over 116,000 local communities in some 210 countries and dependent territories operating under the guidance of a single, elected, global body, the Universal House of Justice. While following common goals and principles for spiritual and social development, each community must adapt itself to the exigencies of its cultural and ecological context. This adaptation is fostered through local, elected, consultative institutions which can develop community resources as well as draw on the national and international resources of the larger community. Each community perceives itself as a "cell" in a "global organism," which itself is a prototype for a future world community.

This relationship between global integration and local adaptation and differentiation is not unlike the relationship between the ecosphere and its component ecosystems.[23] Ecosystems vary greatly according to their locale, but all operate by similar ecological principles and are organically interwoven in the larger encompassing ecosphere. The Baha'i model of an organically structured social order also illustrates how, in general, spiritual and natural principles are correlative.

Aside from the structural arrangements for coordinating global and local concerns, there are several principles outlined in the Baha'i writings that suggest a land and community-based orientation. Agriculture is described as a "fundamental principle" and "village reconstruction" as the initial stage of economic development. Blueprints for the establishment of central community institutions to facilitate community self-reliance and development are also outlined in the Baha'i writings. A key principle is that development should support and benefit whole communities rather than allow individuals or elites to monopolize wealth. Thus the Baha'i view of a global society is one based on individual, family, and local self-reliance, integrated with sophisticated interdependence on the national and global levels.

Male and Female: Equality and Balance

Our prevailing social order is the symbolic expression of the male ego and its tendencies toward rationality and competitiveness. However, qualities of nurturance, intuition, and emotional sensitivity, which many believe to be more associated with the feminine principle, are the qualities most needed in healing our relationship to nature. The emergence of environmental awareness and the equality of women show parallel development.

For Baha'is, the equality of women is seen as an essential objective and a precondition for the establishment of a just and peaceful world. While a full discussion of this important principle is beyond the scope of this paper, the Baha'i writings emphasize that as long as women are prevented from reaching their full potential,

society will be unbalanced. In 1912, 'Abdu'l-Bahá advanced the
following proposition on this important theme:

> man has dominated over woman by reason of his more forceful and
> aggressive qualities both of body and mind. But the scales are already
> shifting—force is losing its weight and mental alertness, intuition, and
> the spiritual qualities of love and service, in which woman is strong,
> are gaining ascendancy. Hence the new age will be an age less mascu-
> line, and more permeated with the feminine ideals.[24]

Summary

The writings of the Baha'i Faith offer a vision of fundamental
global transformation that embodies a new set of principles for
understanding and guiding humanity's relationship to nature.
The religious impulse they contain is a comprehensive source of
spiritual, social, and intellectual resources. They affirm that the
realization of a spiritualized world order which has been the
promise of the sacred scriptures of all ages is now the potential
and requirement of our time.

Elements of this transformative vision include an affirmation
of the divine within creation and an elucidation of the essential
unity of the material and spiritual dimensions throughout the
whole evolutionary process. Humanity, as part of this communion
of life, has gone through a progressive developmental process.
The prevailing social order represents an adolescent stage of this
development. Having passed from the dependence of childhood
through the impetuous autonomy-seeking stage of adolescence,
humanity is now at the point of transition to conscious maturity.
The long historical journey to becoming conscious beings through
separation from nature is culminating in a mature understanding
of life's profound interrelatedness.

The Baha'i writings suggest that this process of maturation requires an expanded religious consciousness complementing and
integral to our scientific advancement. It is only in this context
that the latent capacities of the human spirit can be fully quick-
ened and released. These capacities, such as spirituality, creativity,
and altruism, can encourage selfless action on behalf of the planet,
its peoples, and future generations. They are an infinite resource
in the face of depleting material resources. The development of
spiritual qualities is the requirement of this age and a fruition of
human purpose within the whole evolution of life.

In order to help foster the release of individual spiritual potential and focus it as a force for cultural transformation and moral regeneration, institutions founded on a comprehensive vision of unity need to be established. The Baha'i International Community is itself an embryonic model for such a process of ordered social transformation.

This process of transformation is neither idealistic nor utopian: in the face of the disastrous ecological and human consequences that face us if we continue with "business as usual," this is the new realism. This transformation is possible because the forces that propelled life's evolution from the beginning are still operating within human society. There is reason to believe that the mysterious forces that have "shaped the planet under our feet" and "guided life through its bewildering variety of expression" in natural ecosystems and human cultures "have not suddenly collapsed under the great volume of human affairs in this late twentieth century."[25]

In conclusion, the Baha'i writings offer a vision of wholeness in our relationship to nature and of spiritual purpose in the whole evolution of life that gives a basis for creating a life-affirming culture and empowers individuals to become agents of transformation in developing a sustainable global civilization.

Notes

This essay has been adapted for this issue of *Bucknell Review* from an article that originally appeared in the *Journal of Baha'i Studies* 2, no. 1 (1989) under the title "Spiritual Foundations for an Ecologically Sustainable Society." I would like to acknowledge Paul Hanley, Catherine Fields, and Cyril Boone for their assistance in this adaptation.

1. John Livingston, "Ethics as Prosthetics," in *Environmental Ethics: Philosophical and Policy Perspectives,* ed. Philip P. Hanson (Burnaby: Institute for the Humanities, Simon Fraser University, 1986), 67–81.

2. Thomas Berry, *The Dream of the Earth* (San Francisco: Sierra Club Books, 1988), 36–49.

3. Paul Hanley, "Agriculture: A Fundamental Principle," *Journal of Baha'i Studies* 3, no. 1 (1990): 11–12.

4. Bahá'u'lláh and 'Abdu'l-Bahá, *Baha'i World Faith: Selected Writings of Bahá'u'lláh and 'Abdu'l-Bahá,* rev. ed. (Wilmette, Ill.: Baha'i Publishing Trust, 1956), 364; hereafter, *BWF,* with page references cited in the text.

5. See *Star of the West* 8, no. 11 (27 September 1917): 138.

6. 'Abdu'l-Bahá, *Selections from the Writings of 'Abdu'l-Bahá* (Haifa: Baha'i World Centre, 1978), 27.

7. 'Abdu'l-Bahá, "Tablet to Dr. August Forel," in *The Baha'i Revelation* (London: Baha'i Publishing Trust, 1955), 221–22; hereafter, *BR,* with page references cited in the text.

8. 'Abdu'l-Bahá, *Some Answered Questions,* trans. Laura Clifford Barney, rev. ed. (Wilmette, Ill.: Baha'i Publishing Trust, 1981), 235; hereafter, *SAQ,* with page references cited in the text.

9. 'Abdu'l-Bahá, *Promulgation of Universal Peace,* comp. Howard MacNutt, 2d ed. (Wilmette, Ill.: Baha'i Publishing Trust, 1982), 302.

10. John Hatcher, *The Purpose of Physical Reality* (Wilmette, Ill.: Baha'i Publishing Trust, 1987), 69.

11. Matthew Fox, *Original Blessing: A Primer in Creation Spirituality* (Santa Fe, N.M.: Bear and Co., 1983), 90.

12. Bahá'u'lláh, *Gleanings from the Writings of Bahá'u'lláh,* trans. Shoghi Effendi, 2d ed. (Wilmette, Ill.: Baha'i Publishing Trust, 1976), 166; hereafter, *G,* with page references cited in the text.

13. 'Abdu'l-Bahá, *Makátíb,* 1:436–37. This quote from the Persian book *Makátíb* (unpublished in English) was cited by Bahiyyíh Nakhjavání in her book *Response* (Oxford: Ronald, 1981), 13.

14. Bahá'u'lláh, *Epistle to the Son of the Wolf,* trans. Shoghi Effendi, new ed. (Wilmette, Ill.: Baha'i Publishing Trust, 1988), 44.

15. Henryk Skolimowski, *Eco-Philosophy: Designing New Tactics for Living* (London: Boyars, 1981), 71; hereafter, *E-P,* with page references cited in the text.

16. Though a developmental sequence is suggested, there is no indication of cultural superiority. Different cultural streams have explored and developed different capacities that are all integral to the current period of reconciliation.

17. Nature devoid of spirit became matter, with all sense of its mystery and numinosity lost. Yet the word *matter* has evolved from our original understanding of the earth as "mother." The word for mother in Greek is spelled *meter;* in Latin, *mater* and in Sanskrit, *matr.*

18. William Hatcher, "The Science of Religion," *Baha'i Studies* 2 (1980): 16; hereafter, "Science," with page references cited in the text.

19. Shoghi Effendi, *The World Order of Bahá'u'lláh,* 2d ed. (Wilmette, Ill.: Baha'i Publishing Trust, 1982), 43; hereafter, *WOB,* with page references cited in the text.

20. R. L. Sivard, *World Military and Social Expenditures* (Washington, D.C.: World Priorities, Inc., 1991), 43, and 42.

21. International Institute for Sustainable Development, "Earth Summit Bulletin," no. 2 (1992), 13. Also see World Commission on Environment and Development, *Our Common Future* (Oxford: Oxford University Press, 1987) and "Rio Declaration on Environment and Development" (an Earth Summit press release by the United Nations Department of Public Information, 1992).

22. Bill Devall and George Sessions, *Deep Ecology: Living as if Nature Mattered* (Salt Lake City, Utah: Peregrine Smith Books, 1985), 18.

23. Arthur Lyon Dahl, *Unless and Until: A Baha'i Focus on the Environment* (London: Baha'i Publishing Trust, 1990), 81–82.

24. See *Star of the West* 3, no. 3 (28 April 1912): 4.

25. Berry, *Dream of Earth,* 47.

Hindu Environmentalism: Traditional and Contemporary Resources

Christopher Key Chapple

Loyola Marymount University

THE Hindu religion, perhaps more than any other, is closely linked with a sense of place. It was born on the river banks of the Indus and Ganges rivers, and in recent memory, if a Hindu left the subcontinent to travel across the sea, his or her religious status was compromised. Hinduism holds India's mountains, rivers, and trees sacred, regarding them to be infused with individual spirits *(jīva)* and suffused with an all-pervasive universal consciousness *(brahman)*. Its religious metaphors speak of continuity and reciprocity; its cyclic concept of time does not allow for a fixed moment of creation, nor does it presage a final cataclysmic moment of destruction.

India remains a largely agricultural society. Its products have traditionally posed no great ecological threat. Folk wisdom recognizes the integral nature of waste. The people work to integrate garbage into the cycle of life processes. The lowly is seen as the support for the lofty; both exist in mutual interdependence. People of low status gather the dung of India's omnipresent cow, shaping it into patties that they then sell for cooking fuel. Other people gather scraps of garbage to be reused. Colonies of scavengers surround dump sites, turning offal into treasures. Viewed idyllically, this practice of recycling waste provides an ideal model for ecological living.

A friend of mine once bolted from a bus after a long bumpy ride in India, and blurted to the closest stranger, "Where is the closest latrine?" The stranger, waving his arms expansively, pronounced, "The whole world is a latrine," and sent my friend scurrying to some nearby bushes. Waste in India cannot be separated from landscape; India's mother earth receives and reprocesses all things.

India, with one of the world's fastest growing industrial economies, presents special challenges and resources for the environmentalist. On the one hand, it embraces the comforts of modernity with the passion of 1950s America. On the other hand, it offers a traditional worldview that, if properly applied, can help counter some of the ill effects of the technological trance that has robbed so much from the psyche and physical environment of industrialized nations. To make correlations between traditional worldviews and modern realities is risky business at best, but in this essay I would like to propose that for India and even for the West it might be helpful to tap the conceptual resources of Hinduism for the development of an environmental ethic.

Traditional Resources for Hindu Environmentalism

From the earliest strata of Indian history we find evidence of a city culture attuned to and respectful of natural rhythms; the seals of the Indus valley cities of Mohenjodaro and Harappa depict a meditating figure, thought by some scholars to be the first representation of the god Siva, peacefully surrounded with lush vegetation and peaceful animals. When the Aryan tribes advanced into India from the northwest, they composed songs extolling the powers and wonders of the earth. The Rig Veda lavishes high praise upon various natural forces, regarding them to be divinities worthy of worship. The rivers (Gaṅgā, Yamunā, Sarasvatī, Sindhu) and the earth (Pṛthivī) are regarded to be goddesses, while the winds (Maruts) and fire (Agni) are invoked as male deities. From these earliest hymns an image of the human person arises that sees a continuity between the individual and nature in the creation of the cosmos. The Puruṣa Sūkta states:

> The moon was born from his mind;
> His eyes gave birth to the sun;
> Indra and Agni came from his mouth;
> And Vayu [the wind] from his breath was born.
> From his navel the midair rose;
> The sky arose from his head;
> From feet, the earth; from ears, the directions.
> Thus they formed the worlds.[1]

In this passage, an identity or correlation is proclaimed between the external world and the individual human person. This perspective speaks of an intimacy between human beings and their environment. The unknown author likens our thoughts to the moon, our eyes to the sun, and the powers of victory and fire to the human mouth. Our body itself finds its roots in the earth, stretching upward into the sky. In the Judeo-Christian story, God made men and women in his own image; in this primordial tale, which includes no creator god, the forces that compose the universe are also found within the human body.

This sense of continuity with nature is also found in the *Brhadāranyaka Upanisad,* which draws a simile between a tree and a person:

> As a tree of the forest,
> Just so, surely, is man.
> His hairs are leaves,
> His skin the outer bark.
> From his skin blood,
> Sap from the bark flows forth.
> From him when pierced comes forth
> A stream, as from the tree when struck.
> His pieces of flesh are under-layers of wood.
> The fiber is muscle-like, strong.
> The bones are wood within.
> The marrow is made resembling pith.[2]

Both these images of the human body establish a strong kinship between the self and the world of nature, indicating a worldview that holds inherent respect for the nonhuman realms of existence. The Atharva Veda, the source of India's traditional medicine known as Ayurveda, includes passages in praise of the earth, asking for her beneficence and pledging protection in return. The text states: "The earth is the mother, and I the son of the earth!" (12:12).[3] The author attributes all wealth to the earth and appeals for her to be generous:

> The earth that holds treasures manifold in secret places, wealth, jewels, and gold shall she give to me; she that bestows wealth liberally, the kindly goddess, wealth shall she bestow upon us! (12:44)

Despite this yearning to benefit from her largesse, the author nonetheless harbors a desire not to hurt her, stating, "What, O earth, I dig out of thee, quickly shall that grow again: may I not, O pure one, pierce thy vital spot, (and) not thy heart!" (12:35). This plea indicates a sense of respect and care for the earth, expressing concern that the earth be made aware that the speaker will not hurt her. In yet another verse, the speaker states:

> Thy snowy mountain heights, and thy forests, O earth, shall be kind to us! The brown, the black, the red, the multi-coloured, the firm earth, that is protected by Indra, I have settled upon, not suppressed, not slain, not wounded. (12:11)

In recognition of mother earth's abundance, the Atharva Veda offers both praise of her power and assurances that she will not be harmed by human intervention.

The Hindu tradition, which finds its primary authority in the Vedic literature composed by the Indo-European invaders, offers diverse conceptual resources that can be used to promote ecological sensitivity. In the Vedic hymns, we have seen an intimate relationship between nature and the human order as through various personifications of the earth, water, thunderstorms, and so forth. The Vedic rituals, many of which are still performed today, invoke elemental forces. The *Sāṃkya* tradition reveres the five great elements *(mahābhūta)* of earth, water, fire, air, and space as the essential building blocks of physical reality. From the Upanishads and later Vedantic formulations, all things with form *(saguna)* are seen to be essentially not different from the universal consciousness, or ultimate reality; any thing with form can be an occasion to remember that Brahman which is beyond form *(nirguna)*. In the monistic Hindu model, the human order is seen as an extension of and utterly reliant upon the natural order. In the language of Vedanta, the Brahman is inseparable from its individual manifestations. As stated in the Bhagavad Gita, the person of knowledge "sees no difference between a learned Brahmin, a cow, an elephant, a dog, or an outcast."[4] There is no fundamental difference between ourselves and others; both are undergirded by the common substrate known as Brahman. To violate another creature is to violate Brahman itself. This ethos signals deep concern for harmony among life forms and can be used to advocate a minimal consumption of natural resources.

Throughout the Hindu religious tradition, rituals and ceremonies celebrate myriad manifestations of nature. Pilgrimages abound wherein pious folk trek to sacred places such as mountain tops and the confluence of rivers. Daily observances include the veneration of the four elements and the recitation of ancient hymns that extol their power. Families often do not partake of a meal until food has been shared with birds, who fly through the open windows to receive their offerings. Special trees are placed "under worship," such as the Bodhi Tree under which the Buddha achieved his enlightenment, and the Tulsi tree, a species beloved by and closely associated with the worship of Krishna. In these and countless other examples, a continuity is assumed between the human world and that of nature.

Gandhian Economics

Mahatma Gandhi, who drew upon the existing tradition as learned primarily from Jainas (with input from Tolstoy and British vegetarians) to develop a campaign for national independence, included in some of his writings the basis for what could be interpreted as an environmentally friendly system of economics. He proposed the revitalization of village economies, based on the principles of nonviolence (ahiṃsā) and nonpossession (aparigraha). The purpose of Gandhi's campaign was to make villages self-sufficient, able to cooperate through mutual trade without the importation of foreign produced goods. His means to achieve this was through spinning and weaving cloth and revitalizing other crafts within each village, requiring that schools include these skills as part of the curriculum. Although the aim of this program was to subvert the colonial economic dependence thrust upon India by the British, and despite the fact that Gandhi himself did not object to industrialization, the conservation of energy inherent in the system could be effectively used to counter some of the environmental peril posed in the postmodern world.

Gandhi warned that the urban model associated with industrialization would be detrimental to India's well-being:

> Industrialization on a mass scale will necessarily lead to passive or active exploitation of the villages as the problems of competition and marketing come in. Therefore, we have to concentrate on the village being self-contained.[5]

Gandhi envisioned a network of largely self-sufficient villages, where goods would be both locally produced and consumed:

> My idea of village swaraj is that it is a complete republic, independent of its neighbors for its own vital wants, and yet inter-dependent for many others in which dependence is a necessity. Thus, every village's

first concern will be to grow its own food crops and cotton for its clothes. . . . My economic creed is a complete taboo in respect to all foreign commodities, whose importation is likely to prove harmful to our indigenous interests. This means that we may not in any circumstances import a commodity that can be adequately supplied from our country. (VR, 30)

Though trade policies have recently been changed in India to allow greater foreign investment, for the most part, particularly in regard to food and clothing, contemporary India has abided by Gandhi's precept and imports very few basic commodities.

Using an ironically prophetic voice, Gandhi urges that the people of India remember the sacredness of their land and resist the ills of urban life:

This land of ours was once, we are told, the abode of the Gods. It is not possible to conceive Gods inhabiting a land which is made hideous by the smoke and the din of mill chimneys and factories, and whose roadways are traversed by rushing engines, dragging numerous cars crowded with men who know not for the most part what they are after, who are often absent-minded and whose tempers do not improve by being uncomfortably packed like sardines in boxes and finding themselves in the midst of utter strangers who would oust them if they could and whom they would, in their turn, oust similarly. I refer to these things because they are held to be symbolic of material progress. But they add not an atom to our happiness. (VR, 34)

By minimizing one's needs and the means used to produce those needs, life, according to Gandhi, holds more potential for happiness. In contemporary times, such a lifestyle is an integral part of environmentalism.

Contemporary Hindu Environmentalism

During a trip to the Indian state of Kerala in 1981, I was stunned by the rampant pollution caused by the chemical manufacturers. We had spent a couple of hours traveling by bus to the birthplace of Sankaracharya, one of the great medieval Hindu philosophers, when, all of a sudden, the lush, verdant landscape gave way to brown stubble and the sweet tropical air became foul with chemical stench, revealing the worst of what one encounters in New Jersey or Niagara Falls. We approached the plant—ironically a producer of fertilizer—and were "welcomed" by a dour crowd of employees, who seemed none too thrilled with their factory-dominated existence. As we pulled away from the plant, I commented to one of our hosts that the pollution seemed extreme. He blithely replied that fortunately no one lived nearby. However, as we rounded a bend, a construction project was underway, well within the umbrella of foul air: new housing for factory workers. As we left the general locale and returned to the world of rice paddies, rickshaws, and coconuts, I could not help but fear for the long-term effects of apparently unregulated industry. A couple of years later, sadly, my sense of foreboding was confirmed: the Union Carbide disaster in Bhopal in 1984 took the lives of over 3800 persons in central India.

During this trip, I was struck with the violence to the environment incurred in India by its burgeoning industrialization. The level of awareness in this regard was close to nonexistent, much to my consternation and distress. However, since that time, a noticeable change has taken place: virtually every newspaper routinely includes a story of environmental interest nearly every day. Since the Bhopal Union Carbide disaster, two major centers have been established to serve as clearinghouses for environmental issues. In New Delhi, the Centre for Science and Environment, in addition to other activities, provides a news service that supplies India's many newspapers with stories of environmental and ecological interest. In Ahmedabad, the Centre for Environmental Education, located at the Nehru Foundation for Development, established in 1984, conducts an array of programs designed to enhance the general public's awareness of environmental issues. It offers workshops and produces materials that reach over ten thousand teachers per year. It operates a "News and Features Service" similar to that of the Centre for Science and Environment. It has initiated a rural education program to help stem the destruction of India's remaining forests. It conducts various urban programs, including the promotion of smokeless *chulha*, or woodburning stove, for cooking.[6] In 1986 it launched the Ganga Pollution Awareness Programme and in cooperation with the School of Forestry of the State University of New York, located in Syracuse, it produced a series of environmental films for children.[7]

Many movements in India have taken direct action in an effort to bring attention to environmental concerns. The Chipko movement in Uttar Pradesh involves local women saving trees by embracing them, staving off bulldozers.[8] In traditional Indian folk culture, the tree is sacred and intimately tied to survival of the land and its people. Vandana Shiva, a defender and spokesperson for the Chipko movement, tells the following story in this regard:

> Chipko started in the Himalayas . . . the source of the Ganges River. . . . The Ganges was a mother goddess, and here were prayers for her to be brought to earth. . . . She couldn't just come because her power was so strong that if she landed on earth, she would just destroy it. It's really symbolic of the way we get our monsoon rain. It comes so strong, so powerful, that if we don't have forest cover, we get land-slides and floods. So, the god Shiva had to be requested to help in getting the Ganges down to earth. And Shiva laid out his hair, which was very matted, to break the force of the descent of the Ganga. Shiva's hair is basically seen by a lot of us in India as a metaphor for the vegetation and forests of the Himalayas. . . .
> That's the sort of concept people in India have constantly. And so when they see forests cut, they see the god's tresses being violated. When they see the Ganges being dammed, they see their sacred river being violated. When Indian movements live for a long time, they are very heavily based on these sorts of concepts.[9]

The Chipko movement, whose name means "to embrace," traces its origins to 1913, when there was a major movement to protect forest lands. It became revitalized in 1977 with a group of women in the Himalayan region who tied sacred threads and formed chains around trees in order to prevent their harvest. Women who dwell in forested areas have for millennia lived in relationship with trees; the forest provides fodder, fertilizer, food, water, and fuel. The cutting of forests for purposes of monoculture inter-rupts this ecological balance and has caused great devastation in various parts of India. Although men are involved with Chipko, trees and their protection have long been identified with women in India. Vandana Shiva notes that in Hinduism:

> All the nature deities are always female, by and large, because all of them are considered *prakrti,* the female principle in Hindu cosmology. Also, all of them are nurturing mothers—the trees feed you, the streams feed you, the land feeds you, and everything that nurtures you is a mother. ("Interview," 27)

She tells the story of one woman who battled the development of a quarry which was proposed at the expense of a vast expanse of forest. The contractor had hired two hundred men to harass the demonstrators, who refused to leave, despite being beaten and pelted with stones. When asked, "What is it in you that gives you all this *shakti* [strength]?", the friend replied:

> "Can you see all this grass growing? We come to cut this grass and every year it grows back. And the power in that grass is the power in me. Do you see these trees growing? They are two hundred years old. Every year we lop these trees to feed our cattle and to keep our children alive, so that the children have milk, and still the trees keep growing and still keep nurturing us, and that *shakti* is in me. See this stream? Every year the rain comes, and it could just run off every year, but these trees stay live long after the rain goes, and they keep feeding us. Clear sparkling water better than anything you get in the cities, and I call it living water. Your water you get in the cities is dead; it comes from a tap. This living water gives me life. And that's my *shakti*." ("Interview," 31)

This anecdote shows the simple beauty of a naturally ecological lifestyle, one that is assaulted by increasing urbanization and Western-style "development."

Vandana Shiva explicitly attacks the premises of third world development projects in her book *Staying Alive,* an eloquent appeal to reverse the drive for world homogenization based on the Western model. In addition to a more standard feminist critique, she also develops a theory of nature rooted in *prakrti* and *shakti,* thus utilizing conceptual resources indigenous to Indian tradition. She points out that development policies often entail a shift from holistic, ecologically sound subsistence farming largely conducted by women to cash-crop farming of one product, often enhanced by technology, that is dominated by men. She refers to this practice as "maldevelopment," stating that "it ruptures the co-operative unity of masculine and feminine, and places man, shorn of the feminine principle, above nature and women, and separated from both . . . Nature and women are turned into passive objects, to be used and exploited for the uncontrolled and uncontrollable desires of alienated man."[10] She contrasts the instantiated immediacy of *prakrti,* which views all things as part of a living continuum, with the deadness of things other than human as perceived in the Cartesian-scientific-technological model, wherein they are regarded only for their potential to be transformed into consumable goods. She criticizes the manipulation of seed technology and the development of inorganic fertilizers as potentially harmful to India's ecosystem.

In many ways the current lifestyle of India contains elements that support an environmental perspective. Most persons live within a short scooter ride or walking distance of work. Foodstuffs consumed by Indians are comprised of grains purchased in bulk from the market and cooked with vegetables procured from traveling greengrocers who push their carts through virtually every neighborhood all day long. Waste is collected and used for fertilizer.

Yet things are now beginning to change and modernize in India, for better and for worse. Inoculations and increased knowledge of sanitation have helped decrease the amount of infectious disease in India. But with modernity comes ironic absurdity. One of the great ecological debates for child-rearing Americans has been whether to use cloth or disposable diapers. The laundering of cloth diapers uses a great deal of water and potentially releases cleaning chemicals into the water system. Disposable diapers use tree and petroleum resources and overwhelm landfills. In my early trips to India, this problem simply did not exist. The hard tile floors of Indian homes allowed for easy cleanup of any toddler indiscretions, and children learned to use the toilet at a young age. However, during recent visits to India, the burgeoning advertising industry has struck gold with one new account. In both New Delhi and Bangalore, one can see towering, crudely painted billboards, strapped to bamboo, announcing "Nude is no longer in style," depicting a plump happy tot sporting a disposable diaper. Other signs of potential and real environmental degradation are beginning to appear. In New Delhi, yogurt is served in a disposable plastic container. Private automobiles have begun to proliferate throughout India. The advent of a consumer economy seems to be eroding the possibility for an ecologically sound form of development. Although industrialization and technologization of the subcontinent are modest by American standards, the sheer numbers of people entering into the middle class make it difficult for the same mistakes of Western development to be avoided. One small example is the automobile: India now produces its own small cars, and increasingly they are owned by individuals. By some accounts (and by personal experience) the Delhi area has perhaps the most polluted air in the world. Yet these newly produced vehicles have no emission controls, nor does there seem to be an interest in lobbying for them.

And yet as the general awareness of environmental ravage increases, the best solution seems to be for each individual to make a change in his or her lifestyle. In America, this requires doing with less. In India, it would require not taking on more. Americans now are educating themselves on how not to use the fabulous technology available to us, from household chemicals to nuclear weapons. One Indian housewife commented to me that although she now owns a refrigerator, she does not really use it, because she still buys her food fresh each day, and that her family would not consider buying a car because public transport is relatively inexpensive and convenient.

In the Hindu tradition, people have not been divorced from the earth: to think of themselves as separate from the ongoing and all-pervasive cycle of life and death would be inconceivable. India now faces the challenges of modernity, technology, consumerism, and technological ravage—in short, buying into a trance state where the world and one's relationship to it becomes estranged and objectified. If India can learn from the lessons of Bhopal and resist the enticements of unbridled industrialization, and follow the advice of those who advocate minimal consumption, then an authentically indigenous environmental ethic can emerge.

Notes

1. Rig Veda 10:190, 13–14, trans. Antonio T. de Nicolás, *Meditations through the Rg Veda: Four Dimensional Man* (New York: Nicolas Hays, 1976).

2. *Bṛhadāranyaka Upaniṣad* 3:9, 28, trans. Robert Ernest Hume, in *The Thirteen Principal Upanishads* (London: Oxford University Press, 1921).

3. *Hymns of the Atharva Veda*, trans. Maurice Bloomfield (1897; reprint, New York: Greenwood Press, 1969), 200; the verses quoted in my following text appear on pp.204, 203, and 200, respectively, in this edition.

4. Bhagavad Gita 5:18, trans. B. Srinivasa Murthy (Long Beach, Calif.: Long Beach Publications, 1985).

5. M. K. Gandhi, *The Village Reconstruction* (Bombay: Bharatiya Vidya Bhavan, 1966), 43; hereafter, *VR*, with page references cited in the text.

6. Kartikeya V. Sarabhai, "Strategy for Environmental Education: An Approach for India," 12, a paper presented at the Annual Conference of the North American Association for Environmental Education, Washington, D.C., 1985.

7. "Centre for Environment Education Annual Report, 1987–88," Nehru Foundation for Development, Ahmedabad.

8. Information on this movement is included in the periodical publication *Worldwide Women in the Environment*, P.O. Box 40885, Washington, D.C. 20016.

9. Ann Spanel, "Interview with Vandana Shiva," *Woman of Power* 9 (1988): 27; hereafter, "Interview," with page references cited in the text.

10. Vandana Shiva, *Staying Alive: Women, Ecology, and Development* (London: Zed Books, 1988), 6.

Toward a Buddhist Ecological Cosmology

Brian Brown
Iona College

CONFRONTED by the severity and scope of planetary degradation, humanity haltingly moves to assume responsibility for the protection of the earth. Yet, the response will prove inadequate if it is motivated merely by the narrow confines of human self-interests. Instead, the formulation of ethical principles that will enhance and maximize the integrity of the whole earth community must be informed by an adequate cosmology. Human commitments to biodiversity, natural habitats, and the preservation of planetary air, waters, and soil will be appropriate and consistent to the degree that they are grounded in an understanding of the universe as a coherent, self-emergent reality. Only when the human species knows the fundamental organic continuity between the universe, the earth, the emergence of life in its rich plenitude, and the evolution of human consciousness can humanity properly know itself and be appropriately guided in its future relationship with the planet. If in the past the human species has assumed a proprietary and exploitive dominance over the natural world, it has largely been a function of a radical ignorance of its own coherence with and derivative status within the unfolding story of the universe.[1] Not until humanity knows its own significance as the self-conscious modality of the universe will it be sufficiently dynamized to assume the decisive changes required to halt the ongoing deterioration of the earth community. A functional cosmology in which the universe as primordial self-expressive reality is as much a psychic-spiritual as well as physical-material process which becomes conscious of itself in human thought is the necessity of the present moment.[2]

Within Mahayana Buddhism, the complementary traditions of the *Tathagatagarbha* and *Alayavijnana* represent a cosmology and correspondent anthropology that is strikingly contemporary. Together, they define a coherent understanding of the Buddha Nature, the Mahayana belief in the inherent potentiality of all animate beings to attain the supreme and perfect enlightenment of Buddhahood. As such, they provide the rationale for and description of the Buddhist path as the process in which individual consciousness is transformed into perfect wisdom. The content of that wisdom is reality as a dynamic totality of mutually interdependent causes and conditions, an integral universe of innumerable, mutually interpenetrating, diverse forms and expressions of "wondrous Being," or Suchness (Tathata).[3]

Such an understanding had been deeply rooted and consistently emphasized from the earliest inception of the Buddhist tradition. The principle of pratityasamutpada, or "the-together-rising-up-of-things,"[4] conveyed the notion that the appearing and standing forth into being, the existence, of any particular thing is a dynamic, collaborative process of many other things. Nothing exists in and of itself, but only as a context of relations, a nexus of factors whose peculiar concatenation alone determines the origin, perpetuation, or cessation of that thing. A line from the Pali canon, revered by all the schools of the Buddhist tradition as an original statement of the enlightened founder himself, pithily formulated the fluid contingency which is the very nature of the phenomenal world:

> This being, that becomes;
> from the arising of this, that arises;
> this not becoming, that does not become;
> from the ceasing of this, that ceases.[5]

In such a universe, any element is the combined shape and apparent form of a specific number of other elements; its unique nature is to have none; its identity can only be defined as the expressive manifestation, the conditioned representation, of those other elements. Thus it was that the Buddha and the Abhidharma school of his followers taught that the world of persons and things were just so many clusters, groupings, or literally "heaps" (skandhas) of five basic psycho-physical elements. Rupa, or material form, is the first and includes the four primary elements of earth, water, fire, and air, as well as the five sense organs and their respective sense objects. The second is vedana, representing feelings, while the third, samjna, refers to all possibilities of perceptual experience. The fourth cluster, samskara, includes all good, bad, or indifferent dispositions, tendencies, volitions, strivings, impulses, and emotions. Finally, the fifth basic element is vijnana, or consciousness, as either pure awareness or the process of ideation and thought.

Through the *skandhas*, early Buddhism identified existence as a thoroughly contextual process: no person or thing is an independent, self-subsisting reality, but comes into being, persists, and deceases as a given function of other factors; life perdures only as a complex aggregation of multiple conditions.

From its origin, then, the Buddhist tradition reflects a conceptual framework rooted in the central intuition of an ecological perspective where nothing exists in autonomous isolation, but where everything is defined as the composite derivative and collaborate synthesis of other elements.

The failure of the human mind to adequately grasp the truth of *pratityasamutpada* remained the consistent concern of Buddhist analysis. Ignorance persisted on the one hand in the projection of the ego as the discrete, self-consistent, self-individuating, and self-directing center and end of the individual personality. On the other hand, it manifested a tenacious belief in the autonomous status and independent sufficiency of all other entities or things. The painful alienation *(dukha)* between oneself and the world of persons and things is a function of that primordial ignorance which imputes a false self-derived and self-contained identity to persons and things.

The object of Buddhist soteriology was to bring that ignorance to an end. Through philosophical analysis and meditative wisdom the tradition never departed from its goal of exposing the radically contextual nature of reality, exposing the component parts, the heap of relations that alone give a thing its identity. A striking example of the relentless focus applied by Buddhism to reveal the mutual interdependence and combined aggregation that defines the existence of all phenomena is the *Path of Purification (Visuddhimagga)* by the fifth-century monk, Bhadantacariya Buddhaghosa. One of the most influential scholastic commentaries, exhaustively detailing the types and methods of meditational praxis, the text intensively discloses the feature common to its otherwise various subjects. Specifically, the manual contains innumerable references to, and precise instructions for, meditations on the inevitability and experience of old age, sickness, and death; on the subdivision of the human body into thirty-two parts, each with a specific function and relationship to the others; meditation on varieties of physical decomposition and decay; on the minute details of breathing and eating; and a comprehensive correlation of each of the thirty-two parts of the body (both human and nonhuman) with one of the four primary elements of air, earth, fire, and water. But whether the meditations involve the macabre concentration upon a bloated and festering corpse or the more refined attention to the inflow and outflow of breath, all such exercises share a common purpose: to see reality as it is, namely as a realm in which nothing arises and stands forth into being of its own power, but whose origin and persistence is a function of conditions, factors which are themselves products of other factors. Unifying the rather peculiar and at times exotic meditations is the universality of organic process. Whether it be the process of breathing, the process of age, disease, and dying, or the processes of decomposition and decay, the *Visuddhimagga's* unremitting exposure of phenomena as organic aggregations of multiple constituent elements is designed to pierce the illusion of a world populated by autonomous beings and entities, extraneous and unrelated.

With the subsequent emergence of the Mahayana tradition and the elaboration of the complementary notions of the *Tathagatagarbha* and *Alayavijnana*,[6] the ecologically sophisticated description of reality in the principle of *pratityasamutpada* assumed the status of a coherent cosmology. The earlier Hinayana tradition had identified the precise delineations of phenomenal reality as contingent and dependently co-arisen. But while the intense reductive analysis of persons and things into their clusters of component elements *(skandhas)* accurately reflected the web of multiple conditions which together define the identity of any particular phenomenon, the tradition never addressed the universe as a cohesive, unified reality.

The focus was individual liberation of the mind from the ignorance which projected an illusory significance onto persons and things as absolute, unconditional realities in and for themselves. Freedom from the suffering and unhappiness *(dukha)*, rooted in the subsequent attachments to those erroneously conceived phenomena, was the goal of the path.

With the evolution of the Mahayana schools, Buddhist reflection matured to a more expansive interpretation of the path and the nature of wisdom revealing the truth of *pratityasamutpada*. Incorporating the doctrine of Emptiness *(Sunyata)* of the Madhyamika and Prajnaparamita traditions, the *Ratnagotravibhaga*[7] became the authoritative source for the theory of the *Tathagatagarbha*. Translated as "the embryo of the Tathagata," the term signifies the inherent capacity of all animate beings to attain the supreme and perfect enlightenment of Buddhahood; all beings are embryonic Buddhas *(Tathagatas)* by virtue of their innate endowment with the *Tathagatagarbha.*

An earlier sutra, traditionally credited as introducing the theory of the *Tathagatagarbha,*[8] had defined it as a beginningless; uncreated; unborn; undying; permanent; steadfast; intrinsically pure reality which, when liberated from the defilements of ignorance that conceal it, becomes manifest as the cosmic body of the Buddha *(Dharmakaya);* put otherwise, the cosmic body of the Buddha *(Dharmakaya)* is referred to by the term *Tathagatagarbha* when it remains obscured by ignorance.[9] The implication of this identification of *Tathagatagarbha* as *Dharmakaya* is critical for articulating an adequate contemporary cosmology from within the resources of the Buddhist tradition. While enhancing the role of human consciousness, primary subjectivity is now understood as grounded in the universe itself in its religious symbolization as the cosmic body of the Buddha, the *Dharmakaya.* The Buddhist path could now be interpreted as more than the mere individual struggle to overcome erroneous misconceptions and extricate oneself from the pains of ensuing attachments. With the theory of the *Tathagatagarbha,* the path assumed its macrophase significance while simultaneously intensifying the value of its earlier microphase dimension. The universe, religiously conceived as the cosmic body of the Buddha (the *Dharmakaya*), journeys to perfect self-consciousness as that totality, in and through the human mind. The progressive insights of the human mind into the nature of reality are the embryonic maturations in ever more exact self-awareness of that cosmic body (the *Dharmakaya*).

Without changing that basic cosmology, the *Ratnagotravibhaga* specified the ontological identity of the *Tathagatagarbha* and the *Dharmakaya* as but variant modalities of one and the same unconditional, indeterminate, all-inclusive and nondifferentiating "wondrous Being," or Suchness *(Tathata).* The designations of *Tathagatagarbha* and *Dharmakaya* are merely linguistic distinctions referring to *Tathata* as ultimate reality. When *Tathata* is fully self-conscious of its own integral totality as the primordially pure, immaculate essence *(dhatu)* of all things, is perfectly self-aware as universal body, it is referred to as the *Dharmakaya.* Until it attains that ultimate self-disclosure, *Tathata* is fully present in all sentient beings (as the *Tathagatagarbha*) in various embryonic stages of self-realizations. The movement which characterizes *Tathata* from *Tathagatagarbha* to *Dharmakaya* is the necessary emergence of itself to itself in perfect self-knowledge. Indeed, the *Ratnagotravibhaga* characterizes *Tathata* through *Cittaprakrti* as the innately pure Mind present in all animate beings through which it recognizes itself as the wholeness of reality in the plurality of its forms.

The wisdom which perfects that ultimate self-recognition is nothing other than the truth of *pratityasamutpada* now informed by the doctrine of nonsubstantiality, or Emptiness *(Sunyata)*. Here the Mahayana tradition is consistent with the original insight of and scholastic development by the Hinayana tradition into the nature of phenomena as dependently derived and conditionally produced as the expression of multiple, interdepending factors. But the Mahayana doesn't uncritically assume but creatively incorporates and significantly nuances the earlier articulation. In advancing its reflection on "wondrous Being," or Suchness *(Tathata)*, the *Ratnagotravibhaga* reviews different classes of human beings whose respective insights into the nature of reality represent the acuity with which *Tathata* as the innately pure Mind *(Cittaprakrti)* moves from embryonic self-awareness (i.e., as the *Tathagatagarbha*) to perfect self-consciousness as the essential nature of all things as one cosmic body (i.e., as the *Dharmakaya*).[10]

Beginning with "ordinary beings" whose crass materialism seizes upon persons and things as independent, discrete, self-subsisting entities, *Tathata*'s self-understanding is utterly opaque. Without any clue to the conditional status of phenomena (as constituted by the *skandhas*), such persons define themselves in terms of substantial egohood *(ahamkara)* and their relation to other persons and things is largely a function of their craving and possessive self-reference, i.e., their sense of "mine" *(mamakara)*. Since those ordinary beings lack any sensitivity to the relative, determined and conditional status of phenomena, the notion of nonsubstantiality, or Emptiness *(Sunyata)*, is scarcely conceivable. Among persons with such a degree of ignorance, *Tathata* as *Cittaprakrti* remains fundamentally obscure to itself.

Turning to the classical position of the Hinayana understanding of the doctrine of *pratityasamutpada* as discussed above, the *Ratnagotravibhaga* credits its analytic reflection on and critical awareness of phenomena as dependent and provisional. Differing from the gross superficiality of ordinary beings, representatives of the Hinayana (the *sravakas* and *pratyekabuddhas*) attained a genuine perception into the truth of reality. As fundamentally qualified by a whole series of causes and conditions, persons *(pudgalas)* and things *(dharmas)* in and of themselves were correctly understood by the Hinayana adherents as totally lacking *(sunya)* the perma-

nence and substantiality accorded them by the majority of ignorant persons.

But despite their initial success in overcoming the illusion of
the gross substantiality of existent elements, the Hinayana adepts
became entrapped by the very categories of their analysis. Having
reduced phenomena to their major classifications of the five heaps
(skandhas), the twelve sense-fields *(ayatanas)*, and the eighteen elements *(dhatus)*, they unequivocally devalued phenomenal reality
as essentially marked by impermanence *(anitya)*, suffering *(dukha)*,
absence of ego *(anatma)* and impurity *(asubha)* and regarded it as
a repulsive source of pain and sorrow. Initially more sophisticated
and accurate in its insight into the nature of phenomena as derivative and dependent upon multiple constitutive factors, the Hinayana erred by denigrating the conditionality and relativity of
existence as itself unconditional and absolute. By absolutizing the
classifications of its own analysis (i.e., the *skandhas, ayatanas* and
dhatus) and its consequent descriptions of phenomena (as *anitya,
dukha, anatma,* and *asubha*) into ultimate facts, the Hinayana tradition never perfected the intuition of universal nonsubstantiality,
or Emptiness *(Sunyata)*. In such persons, the self-comprehension
of *Tathata* as the originally pure, undivided essential nature of
phenomenal reality is aborted. Blocked by an ignorance which
fragmented existence into certain fundamental, irreducible units,
Tathata never conceives of itself as the undifferentiated coherence
of the universal whole.

Turning to its own tradition, the *Ratnagotravibhaga* censures the
ignorance of certain Bodhisattvas who are novices to the Mahayana. Unlike the "ordinary beings" and the followers of the Hinayana path, these Bodhisattvas formally acknowledge, though
incorrectly understand, the doctrine of nonsubstantiality *(Sunyata)*. Among this group are those who misapprehend *Sunyata* as
some unconditional reality, transcendent and separate from the
realm of conditioned phenomena. Reified as something to be attained outside of and beyond mundane reality, such a conception
of *Sunyata* implies the denigration of phenomenal existence. Misunderstood and clung to as a reality existing absolutely and inde-
pendently of the five *skandhas* and the entire conditioned world
that is coextensive with them, such a *Sunyata* becomes yet another
expression of ignorance.

An even more serious delusion occurs when the Mahayana doctrine of *Sunyata* is misapprehended as signifying utter nihilism.
Assuming the emptiness of *Sunyata* to mean the actual unreality
of phenomena, dismissing their appearance as the mere products

of an illusory imagination is a perverse distortion of the revelatory nature of *Sunyata*. When images are applied by the Prajnaparamita literature to the emptiness of phenomena as stars; magical apparitions; clouds; dew drops; bubbles; lightning flashes, or reflections of the moon in water, by no means do they postulate the absolute nonexistence of those things.[11] The purpose of those similes is only to deny the status of phenomena as independent, self-subsisting entities; the similes are comparative statements indicating a certain degree of reality and are not unqualified assertions of a total nullity. Rather than deny their existence, the emptiness implied by such images reveals the reality of phenomena as opposed to how they are perceived by the ignorant.

Like stars, things, appearing as so many independent, ultimate realities, are distant, unreachable, unattainable, insignificant, and seen only in the darkness of ignorance; like magical apparitions, their semblance of individual, ultimate significance is a deception and the fraudulent pretense of ignorance; like dew drops, their existence is temporary and evanescent; like bubbles, the factors of experience, while actual, are insubstantial, and lasting but a moment; they are like a flash of lightning and as impermanent as clouds.

By disclosing the emptiness of an independent self-subsistence in all *dharmas*, *Sunyata* doesn't imply the absolute nullity, or nonexistence, of things. As the true nature of phenomena, *Sunyata* doesn't diminish the value of anything, but is the very mode by which their essential nature as a mutually interdependent, cooriginating whole becomes manifest.

In the mind of the more mature Bodhisattva, skillfully avoiding the errors of conceiving *Sunyata* as some ultimate reality existing independent from and transcendent to phenomenal existence, or as suggesting a total nothingness, *Tathata* attains a precise self-awareness. As Mind innately radiant *(Cittaprakrti)*, *Tathata* becomes actually so in the Bodhisattva who, knowing all things as empty of all specific characteristics or determinations that define them as essentially distinct and separate particularities-in-themselves, understands their coherence and totality as one cosmic body, the *Dharmakaya*.

The *Ratnagotravibhaga*'s clarification of *Tathata* in its dual modalities as *Tathagatagarbha* and *Dharmakaya* is important to a Buddhist ecology. It was initially noted above that an adequate environmental ethic must be grounded upon a cosmology capable of rendering the universe as a coherent whole in which human consciousness is an intrinsic self-expression of that larger reality. Human concern for and protection of the earth community will be more carefully informed and appropriately guided when human consciousness comprehends its own significance as evolved from and dependent upon the entire cosmic process. That the universe may understand its entirety in its innumerable particularities defines a clear purpose and singular responsibility for human thought and behavior. Such a cosmology and attendant ethic is indicated by the *Ratnagotravibhaga*'s general analysis of *Tathata*. In the text, Buddhism suggests that "wondrous Being," or Suchness, is the movement toward its own self-revelation. It must come to recognize itself as the essential nature of all things. It can do so in and through the human mind which, grounded upon and informed by *Cittaprakrti* (the noetic determination of *Tathata* and an alternate designation of the *Tathagatagarbha*), attains an ever more exact insight into the nature of reality. From the gross materialism of "ordinary beings" through the more refined analysis into conditional relativity of the Hinayana tradition, past the mistaken notions of Emptiness *(Sunyata)* of some within the Mahayana, the inherent tendency of *Tathata* to know itself as the perfectly pure essence, the Suchness of all things, embryonically moves toward perfect self-realization as one universal reality, or *Dharmakaya*.

The notion of reality as a self-reflecting whole and the status of human consciousness as intrinsic to that process became all the more defined in the Vijnanavada, or "Consciousness Only," tradition. The *Ratnagotravibhaga* delineated *Tathata* as the universal, immaculate essence of phenomenal existence which as embryonically present in all animate beings is referred to as the *Tathagatagarbha*. That the nature of *Tathata* is to determine itself as perfect wisdom and recognize itself in the coherence of its universal integrity was indicated by *Cittaprakrti* as a cognate expression of *Tathagatagarbha*. That designation became explicit in the *Lankavatara Sutra*'s identification of the *Tathagatagarbha* as the *Alayavijnana*, or Absolute Consciousness.[12] The *Lankavatara Sutra* in turn became a critical source for the development of the Vijnanavada as exemplified for the present essay by the *Ch'eng Wei-Shih Lun* of Hsuan Tsang.[13]

According to Hsuan Tsang's treatise, the universe in the plurality of its forms is the self-manifestation of the *Alayavijnana* (*Tathata* as Absolute Consciousness). More specifically, the *Alayavijnana* contains universal seeds *(bijas)* which, as archetypal self-determinations, are actively and persistently projected by the *Alayavijnana* as the innumerable forms of the phenomenal universe; the physical shapes and contours of the cosmos are in fact the universal self-particularizations of consciousness. The apparent solidity and uniform stability of those forms by no means invalidates their origin in, and persistence as, mere consciousness. The abiding character of matter attests to the uninterrupted continuity of the *Alayavijnana*'s self-manifestation. The "consciousness only" of the Vijnanavadin tradition doesn't impugn physical consistency and concrete tangibility. Because it is ideal, does not mean that the empirical world is subject to no laws; idealism is not to be construed as the negation of precise and rigorous spatial-temporal determinations. Instead, they are the very forms in which Absolute Consciousness (the *Alayavijnana*) manifests itself. It is not the material solidity of empirical phenomena, but only the notion or idea of their externality (apart from consciousness) that is disputed by the doctrine of "consciousness only."[14] The univeral *bijas* as the innate self-determinations of the *Alayavijnana* are actively and persistently projected by it as the multiple forms of the phenomenal universe. Since the *Alayavijnana* is the seat of the primordial a priori category of objectivity, specified in the general categories of space and time, and since it (the *Alayavijnana*) is the grounding principle of phenomenal consciousness,[15] to perceive those forms is to perceive them as objective.

The error is to misunderstand this fundamental function of the *Alayavijnana* (i.e., the projection and objectification of phenomena) and to interpret the perceived objectivity of things as evidence of their independent self-subsistence. Yet that is what happens. Due to an inherent ignorance, individual phenomenal consciousness regards itself as an independent autonomous ego.[16] While evolving out of and grounded upon the *Alayavijnana*, phenomenal consciousness fails to understand its own derivative status as essentially dependent upon the *Alayavijnana*. Instead of recognizing the *Alayavijnana* as the unconditional reality, the universal Absolute Consciousness, the generic animating principle of all sentient life, the phenomenal mind misapprehends the *Alayavijnana* as the uniquely particular center of its own discrete self-identity (i.e., as an *atman*). This mode of self-delusion *(atmamoha)* is accompanied by a correspondent self-conceit *(atmamana)* and self-love *(atmasneha)* in which the individual considers itself superior and lofty to all others in its possession of a unique selfhood, to which it develops a profound attachment.

That persistent misapprehension of the *Alayavijnana* by the phenomenal consciousness and its consequent distortion of its own identity as an independent, self-subsistent reality in turn pervades its perception of all other persons and things. Its constant self-regard as an autonomous ego instinctively transfers to its apprehension and interpretation of the phenomenal world which is invested by it with a similar degree of self-reality. If the psychophysical organism is considered to be a discrete, self-determining center of unique personal identity (an *atman*), it is so, over and against a plurality of similarly unrelated egos and a world of unconnected, self-standing objects and things *(dharmas)*. This coordinate form of ignorance which interprets the phenomenal universe as constituted by innumerable discrete particularities, independent from one another and from consciousness, represents the Vijnanavada's continuity with the central intuition of *pratityasamutpada* that had animated the entire development of Buddhist thought from its earliest expression. While earlier traditions had identified the dynamic through which all things come into being as derivatively dependent on a host of multiple conditions, the Vijnanavada stressed that their contingent interdependency is rooted more fundamentally in the ultimacy of Absolute Consciousness (the *Alayavijnana*) which projects and sustains the phenomenal universe as its own ideal manifestation and transformation.

But if phenomenal consciousness is dependently originated from and actively sustained by Absolute Consciousness, the reverse is no less true: the *Alayavijnana* attains its plenary self-awareness as the indeterminate, unconditional nature of all things in and through the human mind. Collectively, the forms of the phenomenal universe and of human individuality are the images *(nimitta)*[17] in and through which Absolute Consciousness appears to and recognizes itself. Since the structure of the phenomenal consciousness evolves from immanent, archetypal self-patternings *(bijas)* of the Absolute Consciousness, and since that phenomenal consciousness exists as the differentiated identity of the Absolute Consciousness,[18] the perceptions of the phenomenal consciousness are the perceptions of the *Alayavijnana*.

Thus, in the cosmology expressed in the complementarity of the *Tathagatagarbha* and *Alayavijnana*, the significance of human consciousness is paramount. Even though the human mind has an instinctive tendency to fragment reality into discrete, unrelated particularities of persons and things as noted above, that inherent ignorance is not the essential nature *(svabhava)* nor the essential mode of activity *(akara)* of phenomenal cosciousness. Rather than being an absolute and definitive state, ignorance is but a qualified condition or "an associated mental activity" (caitta) of the human mind.[19] While human consciousness may be originally deluded about the nature of itself and the universe, it is not itself essentially delusive; it may well be the vehicle through which ignorance is manifested and perpetuated, but it is at the same time the very locus within which wisdom realizes its perfection. Just as the structure of human consciousness originates and assumes its form from the innate self-determinations (bijas) of the Absolute Consciousness, so too does the ignorance which accompanies it germinally develop from within the very ground of the Alayavijnana. But concomitant to and simultaneous with the seeds of ignorance, there likewise exist innate seeds of wisdom[20] which actively inform the mind through five stages of progressive illumination.[21]

Moving from the initial "stage of moral provisioning" through the stages of "intensified effort" and "unimpeded penetrating understanding," wisdom embryonically matures, instructing the mind in the true nature of all things as pratityasamutpada: a universe of mutually interdependent coexistences, emerging from and actively sustained as the ideal self-transformations of Absolute Consciousness. Deepening its understanding of the nonsubstantiality (sunyata) of persons and things, the mind refines its comprehension of, and response to, reality as "consciousness only" (vijnaptimatrata) in the stage of "exercising cultivation" through the final stage of "ultimate realization."

In this process, wisdom perfects itself as it transforms phenomenal consciousness in a twofold form. The tenacity of ignorance in its projection of a multiplicity of independent, autonomous entities dissipates through the mature illumination of the Universal Equality Wisdom (Samatajnana) and the Profound Contemplation Wisdom (Pratyaveksanajnana).

What is critical is that human consciousness is a product neither of ignorance nor of wisdom; its natural condition is, rather, the very interplay of their mutual presence. As indicated in the theory of the Tathagatagarbha, Absolute Reality must come to know itself in the totality of its plenitude as the unconditional, indeterminate Suchness of all things. It can do so, because, as Absolute Consciousness (the Alayavijnana), it projects the plurality of the phenomenal universe as its own self-determinations which it then recognizes as itself in and through human consciousness. Thus, the human mind, itself derivative and conditioned by the Alayavijnana, assumes its status as the self-conscious modality of the Absolute. That the Alayavijnana "seeds" the mind with both ignorance and wisdom suggests that phenomenal consciousness is defined as the active interplay between the two. Fundamentally oriented toward and engaged in the understanding of the universe of which it is a part, human consciousness realizes itself in the necessary dialectic between an ignorance which perceives oneself and the plurality of all other persons and things as essentially discrete, self-subsistent realities, and the wisdom which delineates the emptiness and nonsubstantiality *(sunyata)* of all things, comprehending their innumerable mutual interdependencies in their integrity as one universal body (the *Dharmakaya*). This movement of the mind from ignorance to wisdom, from crass materialism to the universe as sacred body, is the very movement of Absolute Consciousness from an implicit to an explicit self-awareness. Such a cosmology, defining the coincidence of human understanding of reality as the self-intuition of that reality, resonates from within the Buddhist tradition with the indications of contemporary physics and biology. It confirms their image of a primary reality that actualizes a concrete self-awareness in human reflection. Together with them it advocates an urgent challenge that humanity free itself from a distorted arrogance and recognize itself as originated in dependence upon a reality more than itself, that it is conditioned by and coexists in dynamic interdependence with all things. Such a cosmology, grounded in universal Emptiness, would reinvigorate the human in an ethic of reflection upon and care for life in its entirety, as the species which can identify the integrity of the whole in the richness of its diverse particularities.

Notes

1. Thomas Berry is the seminal thinker who has interpreted human cultural history and indicated its future development within the larger dynamics of the universe. See, for example, "The New Story" and his other penetrating essays in *The Dream of the Earth* (San Francisco: Sierra Club Books, 1988).

2. See ibid.

3. Generally translated as "Suchness" or "Thusness," *Tathata* has been more recently rendered as "wondrous Being" by Masao Abe in his profoundly instructive collection of essays, *Zen and Western Thought* (Honolulu: University of Hawaii Press, 1985).

4. Routinely translated as "dependent origination" or "conditioned co-production," the term has been rendered more literally and dynamically by Thomas Berry.

5. Majjhima-Nikaya, 2:32; Samyutta-Nikaya, 2:28.

6. The doctrine of the *Tathagatagarbha* is found in the *Srimala Sutra* and elaborately developed in the *Ratnagotravibhaga*. The *Lankavatara Sutra* and the later *Ch'eng Wei-Shih Lun* define and explain the concept of the *Alayavijnana*.

7. Jikido Takasaki, *A Study on the Ratnagotravibhaga (Uttaratantra)*, Serie Orientale Roma, vol. 33 (Rome: Instituto Italiano per il Madio ed Estremo Oriente, 1966).

8. *The Lion's Roar of Queen Sri-Mala: A Buddhist Scripture on the Tathagatagarbha Theory*, trans. Alex Wayman and Hideko Wayman (New York: Columbia University Press, 1974).

9. See ibid., 104–5.

10. See chapters 10 and 11 of the *Ratnagotravibhaga* and Brian Edward Brown, *The Buddha Nature: A Study of the Tathagatagarbha and Alayavijnana* (Delhi: Motilal Banarsidass, 1991), chap. 6.

11. See, for example, the *Vajracchedika Prajnaparamita*, in *Buddhist Wisdom Books*, trans. Conze (New York: Harper & Row, 1972), 68. See also, Brown, *The Buddha Nature*, 150–51.

12. See Brown, *The Buddha Nature*, 179–81.

13. Hsuan Tsang, *Ch'eng Wei-Shih Lun: The Doctrine of Mere-Consciousness*, trans. Wei Tat (Hong Kong: Ch'eng Wei-Shih Lun Publication Committee, 1973).

14. See Brown, *The Buddha Nature*, 204–5; see also Asok Kumar Chatterjee, *The Yogacara Idealism* (Delhi: Motilal Banarsidass, 1975), 74–75.

15. According to the Vijnanavadin tradition, human consciousness consists of a sevenfold modality. The first five sensorial consciousnesses of seeing, hearing, smelling, tasting, and touching represent the simple awareness of the respective data appearing before consciousness. It is the sixth, *manovijnana*, or mind consciousness, which is the unifying principle of that raw sense information as apprehended by the first five. It accounts for the constitution of objects within consciousness and their intelligibility or rationality. As the consciousness that "perceives ideas," it is the faculty of formal conceptualization. Intellection proper is attributed to the seventh consciousness, the *manas*. It systematically categorizes information and acts upon it, pondering, calculating, and directing means to specific ends. Thus, it is the organ of conative intentionality and the source of ego identity, with its attendant craving, thirst, and desire. All seven modes of consciousness are grounded upon and evolve from the *Alayavijnana*.

16. This form of ignorance, *atmagraha*, is peculiar to the *manas*. See Brown, *The Buddha Nature*, 215ff.

17. For a more detailed explanation of *nimitta* as the self-manifested images of the *Alayavijnana*, see ibid., 217ff.

18. According to the *Ch'eng Wei-Shih Lun*, the *Alayavijnana* and the sevenfold empirical consciousness are said to be simultaneous with and mutually present to each other, and thus, are neither identical to nor different from one another. See Hsuan Tsang, *Ch'eng Wei-Shih Lun*, 131–33.

19. For a clarification of the distinction between *svabhava* and *caitta* as applied to consciousness in the *Lankavatara Sutra* and the *Ch'eng Wei-Shih Lun*, see Brown, *The Buddha Nature*, 223–24.

20. See Hsuan Tsang, *Ch'eng Wei-Shih Lun*, 531–33.

21. For a detailed explanation of the five stages, see ibid., 665–809.

Jainism and Ecology: Views of Nature, Nonviolence, and Vegetarianism

Michael Tobias

He who looketh on creatures, big and small, of the earth, as his own self, comprehendeth this immense universe.[1]

I encountered my first hint of Jainism in a white marble, spotlessly clean temple many years ago while traveling through western India. It was late in the day. Birds were flocking overhead, where a large sculptured spire rose from the center of the temple toward a magenta twilight. As I approached the entrance, a white-robed gentleman quietly surveyed my person, then requested that I be so kind as to leave my watch outside the temple. He explained that leather—the watchband—was not permitted inside the sacred space of Jainism. He used the "ism" rather than referring to the temple itself, and this complicated contraction of the language set me to wondering.

Not long after that, I realized I was a Jain at heart.

There are some seven to ten million Jains today, mostly in India, perhaps thirty thousand in the United States. And they are to be found in many other countries, from Ethiopia to Canada. Until the last century their religion was frequently confused by Western scholars with both Hinduism and Buddhism. But, in fact, it is arguably the oldest living faith in the world, distinct from any other, dating back tens-of-thousands of years. There are incontrovertible references to Jainism as early as the ninth century B.C., three hundred years prior to the emergence of Buddhism.

Unlike the later Brahmanic spiritualist traditions, with their bouquets and contagia of deities, the Jains worship no god. Worship, according to them, is a form of interference, and interference is counter to nature. They revere nature. That is their essential characteristic. The semantics are obtuse, and, as we shall see, contradictory for the Jains themselves. What is the difference between "revere" and "worship" one asks? Perhaps the important distinction rests upon the idea of God, which the Jains dismiss as anthropomorphic, whereas *nature*—the word, the concept, the surreality—necessarily transcends any focal point of conceptualization. At the same time, Jains find themselves neither wishing to interfere with nor willing to accept certain inequities inherent to the natural world. How they go about resolving such dichotomy is the basis of the *Jain ideal.* A community devoted to that ideal, the Jains have worked out their problems in ways that have merged ecology and spirituality: sacrificing self-interest for the greater good, while never losing sight of individual self-worth, which they emphatically uphold.

Conscious love—the striving toward an harmonious coexistence with all beings—is the purposeful, soul-supportive, evolutionary instinct of nature. That, say the Jains, *is* nature. And no semantic penchant, no logical argument, no linguistic or conceptual conquest can do better than that, however it is phrased. Humanity must recognize its place in the natural process. It is not something to be worshiped, not some Other; it is ourselves, in need of nurturance and recognition. Short of that, we are not ourselves. We perish as individuals. And as individuals perish, the entire biological community fragments, endures pain. Pain, say the Jains, is unnecessary. This is a revolutionary notion. It goes on to insist that human beings are like an island of conscience in a sea of turmoil. We have the capability, the responsibility, to protect one another. For the Jains, "one another" means every living organism in the galaxy.

Yet to become a Jain requires far more than mere "reverence" for nature, which is the easy presumption when describing this commonsense orthodoxy. The Jains recognize in their way that reverence is easy, because it is so identifiable with heaven. Anybody can go to heaven. But it takes courage to remain here on earth; to be affirmative and optimistic takes strength: neither to ignore nor forget, but to embrace and conquer. Jainism derives from the word *jina*, which means peaceful conqueror of the self—conqueror of one's inner distractions and temptations. To become a Jain, then, is to embrace this earth *as* heaven, not with any thought of escaping it or eschewing one's responsibility toward it, but in full affirmation of it—every organism, every connection, the whole evolving biosphere.

Every religion assists adepts in an addiction to heaven; every psychotherapy calculates its gain by the notches of paradise, the mental tranquillity, the ideological utopia it can invent. Paradise is easy. Politicians are forever promising it. Great artists in their

passion are invoking it. The Garden of Eden, in other words, is
an idea everywhere about us, in some form or another—from the
Bible to Brueghel, from Club Med to a perfect week in a far
wilderness. Yet the actions that should be concomitant with paradise are rare. Indeed, what should they be? For the Jains, this
earth, with its multitudes of life forms and atoms, is the only true
sphere of meaning, the place of dreams, of action, of moral and
aesthetic culmination. They call it *Jivan-mukhta,* the divine on
earth. But all of these phrases connote an outer admonition,
vague and meandering, that fails to reach the inner soul of
Jainism.

In that inner soul the dimension of thought and behavior can
be simply identified by a word, namely, *ahimsa,* which is Hindi for
nonviolence, or noninterference. Serving the Jain commonality of
purpose, twenty-four exemplary adherents of ahimsa are acknowledged to have achieved the bliss of perfect understanding
and action. The Jains call them *Jinas,* or *Tirthankaras.* They are
not gods, but humans; they did not go to heaven, per se, but to
immortal Earth, their souls richly enshrined somewhere in the
planet's eternal biochemistry. To call it the summit of a sacred
mountain (i.e., Mount Su-Meru or Mount Kailasha in Tibet), or
nirvana, is to cultivate the hieratic inexactitude of yearning beyond
all encapsulation, of language that cannot hope to fix between its
letters the appropriate physical or emotional coordinates. For the
Jains, topography becomes relevant when it has entered the soul.

The most recent of these Jinas, Mahavira, lived in Bihar (eastern India) from 599 to 527 B.C. He was an older contemporary
of Gautama Buddha. Both men shared certain qualities—great
renunciations, extreme psychological embattlement, unfriendly
opponents, legendary hardships. In the case of Mahavira, his ab-
dication of the normal material existence occurred only after his
parents had died. The story goes that he did not want to break
their hearts. After they had passed away, Mahavira took off his
clothes and spent some forty years wandering across India,
preaching the message of peace. He was a total vegetarian and
Jainism itself is adamantly so.

Mahavira's nudity *(acelakka)* is well worth commenting upon,
for it suggests a state of purity and inner unity that is—it must
be acknowledged—rather rare in these times. There are, however,
approximately fifty-six existing naked Jain monks (Digambara sect
Jains) in India who spend but a few days at any one time in any
particular village. They are dependent upon those who will feed
them pure vegetarian food—food that is said to possess but one

sense organ, the sense of touch: specific fruits, vegetables, grains, and nuts. They eat one meal a day, when they are not fasting, food eaten out of the palms of their hands which they consume knowing that even the one-sensed organisms, like soy, want to live, want not to be eaten. All Jains, not just the fewer number of white-robed (Shvetambara) and naked (Digambara) monks, have reduced their consumption; however, they do eat one-sensed beings so as not to starve. A human being, like most animals, has five senses. And because Jains are devoted to minimizing violence on this earth, they recognize that it is better to spare the five-sensed being, even if it means consuming the one-sensed.[2] Such gradations of behavior are consistent with the Jain philosophy of nonabsolutism (anekantavada), the many-sidedness of thinking. What is not suited to this cognitive tolerance, however, is the killing or harming of any organism with more than one sense, except in instances of total self-defense, where once again the minimizing of violence as a general principle is employed.

What this means, practically speaking, is that the Jains have renounced all professions involving harm to animals. Not surprisingly, their ecological avocations have proved to be financially successful and the Jain communities throughout India find themselves economically advantaged. They have used their money philanthropically to perpetuate the practice of ahimsa. They have established animal welfare centers known as panjorapors, compassionate oases in a harsh country where cattle are beloved to death, in essence—left to wander, in other words, until they typically starve to death in old age. Here is an instance of where the Jains will interfere with nature, rescuing the old or infirm animals and caring for them lovingly until they die natural deaths inside the welfare centers. I have seen Jains go to animal markets and purchase sheep and goats destined for the slaughterhouses, rescuing them at any price.

The Jains always granted equal status to women. There was never a caste system among the Jains. How could there be given the Jain conviction that all people can become enlightened? Abortion and contraception are allowed, though abortion itself is not religiously sanctioned. Here again, where the mother's own physical or mental well-being is jeopardized, her adulthood is granted priority status. Pragmatic minimizing of violence is once again at work.

Agricultural professions, timber, even mineral exploitation, most pharmaceutical or any earth-moving enterprises—these are all outside the Jain level of acceptance. Hotels which serve non-vegetarian food to their guests are also against every Jain canon. Jain doctors cannot prescribe any drugs that are derived from animal by-products or were ever tested on animals. Jain lawyers are vehemently opposed to any form of physical punishment. Jains stay out of the military, unless called upon to defend the nation during an active conflict. Jains even forego silk saris, so fundamental to pan-Indian fashion, knowing as they do that approximately ten thousand silk worms are boiled alive to make a single garment. As for Jain monks, they are celibate, but not for the reasons asexuality has been ordained in other religious quarters. For the Jain mendicants, ejaculation is perceived in stark terms: it kills, on average, seventy-five million spermatazoa, while wreaking havoc with the bacterial balance of a woman's genitalia. However, lay Jains propagate themselves despite these uncomfortable recognitions, always oriented in their hearts toward that day when they too can renounce sex, renounce automobiles (cars kill bugs), and simply walk naked, barefoot, throughout their homeland, practicing the primary rules of ahimsa.[3]

Rules are basic to Jain ecology. They translate into daily practices that are meant to inhibit the unrestrained inflow of daily sensation, passion, karma. Karma covers the soul, say the Jains, the way a cataract clouds and inhibits the vision. The goal of the Jain is to restrict and eventually banish the accumulation of karma—material goods, passions, ill-will toward others, complexity, haste, narcissism, ego in all its phases—so that the soul can be eventually free of inconsequential attachments and harmful deeds. They call this condition *kshayika-samyak-darshana,* translated as true insight through the destruction of karma. When that day comes, a Jain will have achieved his *moksha,* or liberation.

Jains have their own form of Ten Commandments (the five *anuvratas,* or vows). These major convictions consist of *ahimsa* (nonviolence—with literally hundreds of psychoanalyzed forms of behavior to avoid, or to embrace), *satya* (truth), *asteya* (not stealing), *brahmacarya* (sexual abstinence), and *aparigraha* (nonpossession). In addition, there are eleven *pratimas*—stages of spiritual progress—and eight *mulagunas,* the basic restraints. In negotiating their way through this labyrinth of injunctions, the lay Jains strive toward monkhood, some more fervently than others. In reality, few achieve that state of complete itinerant renunciation, best articulated, perhaps, by a naked Digambara who once sat with me at a temple above the city of Indore and spoke the following words:

"Twenty-two years ago I took the vow of nudity. Extraordinary as it may appear to you, nudity has become natural to us. . . . We do not possess anything whatsoever and we do not have to tell people to likewise give up their worldly possessions. Our example itself conveys the fact that here is a man who can be happy without having or wanting anything. It is important to see that what hurts himself also must hurt others and what gives happiness to others alone can give happiness to himself. It is ahimsa that makes for friendship between father and son, and love between husband and wife. With these words I bless you. May the whole world remain in peace."[4]

There is in Jainism a practice of "temporary asceticism," much like a fasting or meditation retreat, which makes a monk out of a businessman for a day, or a weekend, or as long as he or she wishes to emulate the foregoing convictions.

I have spent considerable time in India in this mode of impermanent austerity, or *tapas*. But it is a mistake to assume that Asia is where ascetics can best manage. India is not the essential ingredient of such behavior. Jain awareness is what matters. And it is as appropriate in the U.N. General Assembly or the World Court at the Hague as it is on Wall Street or in Hollywood or Washington. The space of this meditation never ceases, never need change: a cafe in downtown Tokyo; the sculptured caverns of ancient Ellora; a marble enclave atop a high Maharastran tropic; a drowsy train headed to nowhere, across Siberia—whatever the personal circumstances of time or place, the same cacophany of senses is rushing in to proclaim the earth itself as the basis for humble and reverential thought and deed. Jainism can work in the industrial sections of Manchester, England or in a place like Ahmedabad, along a sleepy Gujarati river, where Gandhi spent many years spinning his own fabric, meditating, building a case against the British occupation of India, and practicing the Jainism which his earliest mentor, a Jain teacher, and his closest friend—a Jain ascetic—had inculcated in the Mahatma. "If ahimsa be not the law of our being, then my whole argument falls to pieces," wrote Gandhi.

Jainism can work, and must work, anywhere, everywhere. It is ecological shepherding taken to its logical conclusion. Ironically, the Jains within India have become remarkably adept capitalists. Though they collectively account for but one percent of India's nearly one billion residents, they pay a proportionate lion's share of the country's taxes, as well as providing generous philanthropic donations. Their ecological professions have proven to be among

the most lucrative enterprises: businesses like law, computer soft-
ware, publishing, education, diamond cutting, the judiciary,
administration. None of these activities would be considered pure
by the Jain clerics. Diamond cutting, for example, disturbs the
body of the earth, not so much on the finger of one's bride, but
in the mines where the uncut stones are rudely hewn. Publishing
is even more injurious, for the obvious reasons. A single edition
of a Sunday newspaper in Bombay, regardless of the amount of
recycled pulp, has taken its devastating toll on the forests and the
surrounding ecosystem.[5] Nevertheless, by comparison with most
of their contemporaries, the Jains have gone a remarkable distance in minimizing harmful impact on the earth's ecosystems.
And what I was to discover in my various encounters with new
Jain friends—jurists, businessmen, professors, monks, children,
industrialists, doctors, engineers, nuns and monks, photogra-
phers, etc.—is the stunning extent to which they are constantly
talking about and attuned to a meditation on nature, simplifica-
tion, the largesse which is, in their firm estimation, appropriate
to human conscience.[6] According to the Jain approach, the gran-
diose has fused with the humble; an entire living planet has been
incorporated into the germ of a human ideal: the protection of
all life forms.

Concerned with understanding the roots of human aggression
and the possibilities for reversing those tendencies inside a person,
the Jains have qualified their ecological thinking by reference to
the unconscious and to rudimentary psychology; this motif is
everywhere to be found in their approximately forty-six remaining ancient texts, or *Agama,* all written in an ancient Maga-
dhan language known as Ardhamagadhi. Those philosophic and
poetic writings include twelve *Anga* and thirty-four *Angabahya.* By
psychoanalyzing violence, breaking down daily *himsa,* or harm,
into its minute parts, the Jains have discovered the wellspring of
compassion. At every juncture of human behavior they have di-
vined right and wrong, signaling hope, allowing for gentleness,
finding a path toward love that is viable, humble, and appropriate
to everyone. The *dharma-tirtha,* or holy path, is the result of daily
salutations *(namaskara-mantra),* compassion, empathy and charity
(jiva-daya), care in walking *(irya-samiti),* forgiveness *(kshama),* universal friendliness *(maitri),* affirmation *(astikya),* the sharing with
guests *(atithi-samvibhaga),* critical self-examination *(alocana),* constant meditation *(dhyana),* a vast realm of behavioral restraints
(gunavratas), and aversion leading to renunciation *(vairagya).*
These many assertions of a daily quest, taken together, are the

basis for liberation in this life, the realization that all souls are interdependent *(parasparopagraho jīvanam)*.

Mahavira had stated, "All breathing, existing, living, sentient creatures should not be slain, nor treated with violence, nor abused, nor tormented, nor driven away. This is the pure unchangeable external law."[7] The ideal goes far beyond biology. The fact of fickle evolution is no excuse for bad behavior, say the Jains. Evolution does not condemn us to anything. Our choices condemn us. We are not clouds in trousers, drifting out of control, but forces for empathy, capable of adroit and systematic deliberation. As Thoreau cautioned, we must live our lives deliberately, such that the very ecological ground rules that have surfaced in this century are now seen to have had their origins many millennia ago in a cultural and spiritual phenomenon known as Jainism.

For the Jains, ecology and meditation—the inner contemplative faculties, and the outer activism of will and body—are one and the same: an Earth in the Self that becomes, again, the Earth. The concept is a concept, but it is also a revolutionary (though self-evident) form of action—a realm given to the identifying and solving of misgivings, imbalance, trauma, and sadness: ecological activism that is introspective, contemplation that is extroverted.

Jain ecology is thus a process of focusing on nature through the practices of restraint, meditation, and action. This inner attentiveness then blossoms into an embrace of the life principle which is the earth and all of its interrelational beings. That focus, known as *samayika*, relates earth processes to an understanding of the self. The complexity of this fusion, ecologically speaking, can be understood in any number of important ways. For example, when one species of tree goes extinct in the tropics, dozens, possibly even hundreds, of animal species are likely to go extinct with it. We may chop down a tree and say, "In all humility, I have chopped down only one tree to build a simple cabin for myself." Or, we may acknowledge that to chop down that tree is to cause unimaginable harm, to fuel what is understood to be the *tragedy of the commons* syndrome. This is the contradiction inherent to the human psyche. The adversaries are clearly drawn in Jainism. And the soul, the *jīva*, is its battleground. According to the Jains, every soul in every organism, is an individual, with a dream, a want, a fervent hope. All organisms feel pain. No one wants to suffer, neither the bacteria in one's armpit, the 36,000 cubic feet of life in a redwood tree, the tulip, nor the child. We are all individuals, to repeat: beings with souls, beings with needs. And we must be allowed to evolve according to our own inner energies. Thus, the

Jains have sought to protect the wildness in everyone; to reinstate the dignity and original purpose of the wilderness; to reconnect with the nature in everything—but to do so with absolute nonviolence. This imposes a colossal gymnasatic on all thought and behavior. Jainism has undertaken to walk that tightrope. What is remarkable is its delicate balance.

To argue, as many have done, that animals and insects kill one another, an implicit challenge to the high hopes of Jainism, is to ignore the great Jain calling which recognizes violence in nature but vehemently insists that human beings, and other animals as well, have the ability to reverse what is pernicious in the world, to celebrate and coddle, to love and nurture. It does not matter whether one accepts the premise that the animal world is a Hobbesian maelstrom of aggressive genes and self-defense, of hedonistic impulses and only rare instances of kin altruism. For the Jains, inherited or not, it is our responsibility as feeling, thinking beings to make loving the preferred medium of exchange on earth.

Because the Jains have so acutely examined the violence everywhere endemic to nature, they have systematically retreated into communities that are typically densely populated. Jain ascetics do not live in caves far removed from others. They depend upon city and village lay Jains for their food and sustenance. It means that there is no "wilderness tradition" among the Jains, no agricultural or pastoral reveries, no sense of solitary poets and monastic mists. Instead, the Jains are gregarious, living off the land, but doing so according to a pattern of minimal consumption, minimalist impact, and discrete movement and presence. In no other religion or way of life has the ideal been so extreme. But what is striking and uncanny is that ideal's accessibility to individuals.

In one of many such stories, it is said that Mahavira, in a former life, was a lion who—upon speaking with a Jain monk—resolved to die of hunger rather than harm any other living being. And upon his death, was immediately reborn the 24th Tirthankara. We may well pass on without having learned many answers, but the same questions of a life-force with which Jainism is preeminently concerned will always prevail. Such questions concern the universal decency and the possibilities for joy and empathy which are our responsibility to engender as compassionate, rational individuals confronted by a sea of tumultuous evolution. The soul of Jainism is thus about stewardship, requiring human diligence, human conscience, and human love. Jain ecology is nothing more than universal love (mettim bhavehi).[8]

Once that ecological meditation, that Jain summoning of feeling, has begun, there is no turning back.

> "Khamemi sabbajive
> sabbe jiva khamantu me
> metti me sabbabhuyesu
> veram majjha na kenavi"

These words of a Digambara monk were spoken to me in the temple village of Taranga, meaning, "I forgive all beings, may all beings forgive me. I have friendship toward all, malice toward none."[9]

Notes

This essay is an adaptation of the author's piece, "The Soul of Jainism," in Michael Tobias, *Environmental Meditation* (Santa Cruz, Calif.: Crossing Press, 1993).

1. Excerpt from the teachings of the Jinas, in Jyoti Prasad Jain, *Religion and Culture of the Jains* (New Delhi: Bharatiya Jnanpith, 1983), 187.

2. And this minimizing of violence includes one's speech. See, for example, the first *Anga*, the *Acaranga Sutra* (book 2, lecture 4, lesson 2) in *Jaina Sutras*, Part 1, trans. Hermann Jacobi (New Delhi: Motilal Banarsidass, 1980), 152–56: Mahavira has written that monks or nuns must use a sinless language when describing nature. Trees should be described as "magnificent," "noble." When speaking of wild fruits, the monk or nun should never say, "They are ripe, they should be cooked or eaten, they are just in season, or soft, or they have just split," in as much as these are invocations to destroy such fruit. Similarly, with regard to vegetables, Mahavira writes, "A monk or a nun, seeing many vegetables, should not speak about them in this way: 'They are ripe, they are dark coloured, shining, fit to be fried or roasted or eaten." Instead, the monk or nun should say, "They are grown up, they are fully grown, they are strong, they are excellent, they are run to seed, they have spread their seed, they are full of sap." In other words, they should be left alone, to be enjoyed as miracles in and of themselves, free to be souls, free to evolve.

If the monk wishes to "eat or suck one half of a mango or a mango's peel or rind or sap or smaller particles" the monk may only do so if the fruit is free from eggs, is injured, and has been nibbled at already by other animals. The same holds for sugar cane, and nearly every wild fruit or vegetable that is allowed. The very willingness to eat, under circumscribed conditions, hinges strictly upon the monk's recognition that not to eat *something* is suicide, which is also the infliction of pain. Nevertheless, that *something* must be minimal (ibid., lecture 7, lesson 2, pp. 173–77).

The extent to which such ethical delineation proceeds is even to be noted in the twenty-one rules which occasion the wandering monk's defecation. For example, in keeping with current park and wilderness regulations, "a monk or a nun should not ease nature at sacred places near rivers, marshes or ponds, or in a conduit." In relieving themselves, the monk or nun—and by inference, all human beings—must avoid harming "shrubs, vegetables, or roots," and any "places which contain leaves, flowers, fruits, seeds, or sprouts" (ibid., lecture 2, pp. 120–35). In the *Kalpasutra* (117) it is stated that Mahavira "was self-restrained in his way-faring, his speech and his desires, as well as in holding and rightly placing the begging-bowl. He was circumspect in discarding excretia, urine, saliva, phlegm, or body dirt. He was self-controlled in mind, speech, and body. He had restrained his heart, his tongue, his body, his senses, and his carnal desires. He was free of anger, pride, deceit, and greed. His spirit was calm, composed, and tranquil." In essence, "He was liberated from the knots of karma. He had become ego-free and free from all sense of possessiveness . . . He was like a pure bronze vessel emptied of all water."

3. The image conjured by Mahavira is of the most extraordinarily delicate avoidance. "A monk or a nun wandering from village to village should look forward for four cubits, and seeing animals they should move on by walking on his toes or heels or the sides of his feet. If there be some bypath, they should choose it, and not go straight on; then they may circumspectly wander from village to village. A monk or a nun wandering from village to village, on whose way there are living beings, seeds, grass, water, or mud, should not go straight if there be an unobstructed byway; then they may circumspectly wander from village to village" (ibid., lecture 3, lesson 1, p. 137).

4. Translated at the Digambara Temple above Indore in January 1986 by P. S. Jaini, consultant to my PBS film *Ahimsa: Non-Violence* (Los Angeles, Direct Cinema Ltd.).

5. The basic doctrine of Jainism hinges upon this prudent, empathetic avoidance of pain, in oneself, and in the rest of the cosmos. Mahavira wrote that "You are that which you intend to hit, injure, insult, torment, persecute, torture, enslave or kill." This injunction even applies to nonanimate beings. In the *Uttaradhyayana* it is written, "One should not permit [or consent to] the killing of living beings; then he will perhaps be delivered from all misery; thus have spoken the preceptors who have proclaimed the Law of ascetics. . . . In thoughts, words, and acts he should do nothing injurious to beings who people the world, whether they move or not" (lecture 9). Later, in the same *Anga*, it is written, "Having thus learned [the nature of] living beings and lifeless things which is in accordance with the principles of reasoning, and believing in it, a sage should delight in self-control" (lecture 36).

Jains are the first to criticize themselves for their own shortcomings. While their faith is adamant about vegetarianism, Jains are not, as a rule, so-called *vegans*, vegetarians who also abstain from dairy products. Jains do drink milk. Moreover, some wear leather, drink alcohol, eat garlic. Many nonmonks drive cars, fly in airplanes (both activities traditionally forbidden of monks), acquire possessions, build houses, run businesses, and so forth.

6. Among the younger generation of Jains, there is a radical sense that the ancient precepts of their faith have more relevance than ever before. In the *Young Jains International Newsletter*, published out of Middlesex, England, emphasis is increasingly upon such exemplary organizations as Greenpeace, the Red Cross, and Amnesty International, and upon methods of focusing the Jain nonviolent philosophy in the contemporary arena. Writing in the October-December issue of 1992 on the subject of "Jainism in Business and Professional Life," Shashikant Mehta says, "To raise your standard of living you must raise your standard of giving. . . . If you serve others genuinely, the prosperity is bound to come. That is a law of nature." Similarly, Atul Shah has devised a compelling series of "experiments" with Jainism for young people everyday. Instead of a private car, use public transportation. Read, rather than watch television. Wash clothes by hand. Minimize the use of all material possessions for one week. Peace marches, nonviolent study groups, a Jain scientist in Antarctica designing a vegetarian approach to life at a glacial basecamp, the systematic rejection of those aspects of the fashion industry that incorporate animal by-products—from down to cashmere, leather, and fur—and an increasingly outspoken community are all characteristics of a vital evolution in Jainism. Acknowledging this growth, and its underlying tenets, L. M. Singhvi, the author of the *Jain Declaration on Nature* (1990), summarizes by stating that "It is this conception of life and its eternal coherence, in which human beings have an inescapable ethical responsibility, that made the Jain tradition a cradle for the creed of environmental protection and harmony."

7. *Acaranga Sutra* (book 2, lecture 4, lesson 1) in *Jaina Sutras*, 36.

8. See Haripada Chakraborti, *Asceticism in Ancient India* (Calcutta: Punthi Pustak, 1973), 423, 425.

9. Translated by Dr. Padmanabh S. Jaini, University of California, Berkeley, in an unpublished essay, "Ahimsa: A Jaina Way of Personal Discipline," 21.

Ecological Themes in Taoism and Confucianism

Mary Evelyn Tucker

Bucknell University

Humans do not oppose Earth and therefore can comfort all things, for their standard is the Earth. Earth does not oppose Heaven and therefore can sustain all things, for its standard is Heaven. Heaven does not oppose Tao and therefore can cover all things, for its standard is Tao. Tao does not oppose Nature and therefore it attains its character of being. (A Taoist commentary from Wang Pi, 226–249 C.E.)[1]

Mencius answered [King Hui], "If your majesty can practice a humane government to the people, reduce punishments and fines, lower taxes and levies, make it possible for the fields to be plowed deep and the weeding well done, men of strong body, in their days of leisure may cultivate their filial piety, brotherly respect, loyalty, and faithfulness, thereby serving their fathers and elder brothers at home and their elders and superiors abroad." (A Confucian text from Mencius, 372–289 B.C.E.)[2]

NEARLY two decades ago Thomas Berry called for "creating a new consciousness of the multiform religious traditions of humankind" as a means toward renewal of the human spirit in addressing the urgent problems of contemporary society.[3] More recently Tu Wei-ming has written of the need to go "beyond the Enlightenment mentality" in exploring the spiritual resources of the global community to meet the challenge of the ecological crisis.[4]

In drawing upon the great religious traditions of the past for a new ecological orientation in the present, it is clear that the traditions of East Asia have much to offer. My method in this essay is to examine some of the principal texts of Taoism and Confucianism for a phenomenological description of ecological worldviews embedded in these traditions. I risk the inevitable distortions of reducing complex teachings from 2500-year-old traditions to generalizations that need qualification and development. I am also relying primarily on the philosophical and religious ideas of these traditions as evident in their texts and am not discussing their varied religious practices which arose in different periods of Chinese history. Nor am I making claims for a historical consciousness in China of the issues of ecology as we are beginning to understand them in the late twentieth century. Furthermore, I am aware of the ever-present gap between theoretical positions and practical applications in dealing with the environment throughout history.[5] I am also conscious of the dark side of each religious tradition as it developed in particular historical contexts. Nonetheless, in seeking guidance from the past it is becoming increasingly important to examine the perspectives of earlier civilizations and their attitudes toward nature as we seek new and more comprehensive worldviews and environmental ethics in the present.[6] There is not sufficient time or space to work out all of these methodological issues here. However, I would suggest that this project is an important step in creating a new ecumenism of the multiform religious traditions of the human community in dialogue with pressing contemporary problems such as the environment and social justice.[7]

General Comments on Taoism and Confucianism

The two indigenous traditions of China, Taoism and Confucianism, arose in the so-called Axial Age in the first millennium before the birth of Christ. As Karl Jaspers noted, this was approximately the same time as the philosophers in Greece, the prophets in Israel, Zoroaster in Persia, and Buddha in India.[8] In China this period in the Chou dynasty was a time of great intellectual creativity known as the age of the 100 philosophers.

Although there are many historical uncertainties and ongoing scholarly debates about the life and the writings of Lao Tzu and Confucius, it is indisputable that these two figures are of primary importance in Chinese religion and philosophy. Indeed, some writers on Chinese thought see these traditions as complementary to each other and in a kind of creative tension. While Taoism and Confucianism are quite different in their specific teachings, they share a worldview that might be described as organic, vitalistic, and holistic. They see the universe as a dynamic, ongoing process of continual transformation.[9] The creativity and unity of the cosmos are constant themes which appear in the Taoist and Confucian texts. The human has a special role in this vitalistic universe. This is viewed in a more passive manner by the Taoists and a more active mode by the Confucians.

It is, however, this organic, vitalistic worldview which has special relevance for developing a contemporary ecological perspective. Indeed, it can be said that within this holistic view Taoist and Confucian thought might provide an important balance of passive and active models for ecological theory and practice.[10] Like a yin-yang circle of complementary opposites, Taoist and Confucian thinkers have evoked important considerations from each other and may still do the same for us today.

In very general terms we can compare and contrast these two traditions as follows. Taoism emphasizes primary causality as resting in the Tao, while Confucianism stresses the importance of secondary causality in the activities of human beings. Thus the principal concern in Taoism is for harmony with the Tao, the nameless Way which is the source of all existence. In Confucianism the stress is on how humans can live together and create a just society with a benevolent government. For both the Taoists and the Confucians harmony with nature is important. The Taoists emphasize the primacy of unmediated closeness to nature to encourage simplicity and spontaneity in individuals and in human relations. For the Taoists, developing techniques of meditation is critical. The Confucians, especially the Neo-Confucians, stress harmonizing with the changing patterns in nature so as to adapt human action and human society appropriately to nature's deeper rhythms. For them the *Book of Changes* is an important means of establishing balance with nature and with other humans.

For the Taoists, in order to be in consonance with the Tao in nature one must withdraw from active involvement in social and political affairs and learn how to preserve and nourish nature and human life. For the Confucians, social and political commitment was an indispensable part of human responsibility to create an orderly society in harmony with nature. Indeed, for the Confucians cultivating oneself morally and intellectually was a means of establishing a peaceful and productive society. The ideal for the Taoist, then, was the hermit in a mountain retreat, while for the Confucian it was the sage, the teacher, and the civil servant in the midst of affairs of government and education. Taoism did provide a model of an ideal ruler, but one who led without overt involvement but rather by subtle indirection and detachment. The Confucians, on the other hand, called for a moral ruler who would be like a pole star for the people, practicing humane government for the benefit of all. The Taoists stressed the principle of non-egocentric action *(wu wei)* in harmony with nature for both ruler and followers. The Confucians, on the other hand, underscored the importance of human action for the betterment of society by the ruler, ministers, teachers, and ordinary citizens. A pristine innocence and spontaneity was valued by the Taoists, while the Confucians continually emphasized humanistic education and ethical practice for the improvement of individuals and society as a whole.

It is perhaps some combination of these two perspectives which may be fruitful for our own thinking today. In order to understand and respect natural processes, we need a greater Taoist attention to the subtle unfolding of the principles and processes of nature. As the deep ecologists constantly remind us, without this fine attunement to the complexities of nature and to ourselves as one species among many others, we may continue to contribute unwittingly to destructive environmental practices. Yet without the Confucian understanding of the importance of moral leadership, an emphasis on education, and a sense of human responsibility to a larger community of life, we may lose the opportunity to change the current pattern of assault on the natural world. Taoism challenges us to radically reexamine human-earth relations, while Confucianism calls us to rethink the profound interconnection of individual-society-nature. Let us turn to examine the worldview of each of these traditions and their potential contributions to a newly emerging environmental ethics.

Taoism and Ecology: Cosmology and Ethics

The principal text of Taoism is the *Tao te Ching (The Way and Its Power)*, also known by the title *Lao Tzu*, its author. There have been numerous translations of this text into many languages and perhaps no other Chinese work compares to it in terms of international popularity.[11] The *Tao te Ching* contains a cosmology and an ethics which may have some relevance in our contemporary discussions on ecology.

In terms of cosmology the Tao refers to the unmanifest source of all life which is eternal and ineffable yet fecund and creative. "The Nameless [the Tao] is the origin of Heaven and Earth; / The named is the mother of all things."[12] The Tao, then, is the self-existent source of all things, namely, a primary cause. It is both a power which generates and a process which sustains. It is the unity behind the multiplicity of the manifest world. It is beyond distinction or name and can only be approached through image, paradox, or intuition. In its manifest form in the phenomenal world it is said to have no particular characteristics and thus be empty. As such it is full of potentiality. Indeed, the "Tao is empty (like a bowl), / It may be used but its capacity is never exhausted" (4). It can be described, however, with images such as valley, womb, and vessel, suggestive of receptivity and productivity.

The implications of this holistic cosmology for an environmental ethic should be somewhat self-evident. There is a distinct emphasis in Taoist thought on valuing nature for its own sake, not for utilitarian ends. The natural world is not a resource to exploit but a complex of dynamic life processes to appreciate and respect. Harmony with nature rather than control is the ultimate Taoist goal. This tradition has certain affinities with contemporary movements in deep ecology which decry an overly anthropocentric position of human dominance over nature.[13] Indeed, the Taoists, like the deep ecologists, would say that manipulation of nature will only lead to counterproductive results.[14]

To achieve harmony with nature the Taoists value simplicity and spontaneity. They distrust education and the imposition of moral standards as interfering with true naturalness. Intuitive knowledge and a pristine innocence are highly regarded.[15] A direct, unmediated encounter with nature is far better than book knowledge or hypocritical morality. As Lao Tzu urges, one should: "Abandon sageliness and discard wisdom; Then the people will benefit a hundredfold. . . . Manifest plainness, embrace simplicity, reduce selfishness, have few desires" (19).

Moreover, in terms of human action that which is understated, not forceful or directive, is considered optimal. Excess, extravagance, and arrogance are to be avoided. Nonegocentric action (wu wei) which is free from desire and attachments is essential.[16] In short, "By acting without action, all things will be in order" (3). In light of this, the Tao te Ching celebrates the paradox that yielding brings strength, passivity creates power, death creates new life.

These ideas are illustrated in the text with feminine images of fecundity and strength springing from openness and receptivity such as in motherhood, in an empty vessel, or in a valley. They also underlie images such as water wearing away at solid rock or the idea of an uncarved block waiting to reveal its form at the hands of a skilled sculptor. These demonstrate the potentiality and generative power which exist in unexpected and hidden places.[17]

> He who knows the male (active force) and keeps to the
>   female (the passive force or receptive element)
> Becomes the ravine of the world.
> .   .   .   .
> He will never part from eternal virtue.
> He who knows glory but keeps to humility,
> Becomes the valley of the world.
> .   .   .   .
> He will be proficient in eternal virtue,
> And returns to the state of simplicity (uncarved wood).
>
> (28)

In short, the *Tao te Ching* demonstrates the ultimate paradox of the coincidence of opposites, namely, that yielding is a form of strength. (This is clearly illustrated in the martial art of judo, which means the "way of yielding.") Indeed, the lesson of Taoism is that reversal is the movement of the Tao, for things easily turn into their opposites. "Reversion is the action of Tao. Weakness is the function of Tao. All things in the world come from being. And being comes from non-being" (40).

Thus both personally and politically Taoism calls for noninterfering action. A Taoist government would be one of conscious detachment and the ideal leader would be one who governs least. While this seems antithetical to the Confucian notion of active political involvement, the wisdom of the Taoist ideal of noninterference was not lost in the highest quarters of Chinese Confucian government. Over one of the thrones in the imperial palace in Beijing are the characters for *wu wei* (nonegocentric action), perhaps serving as a reminder of the importance of a detached attitude in political affairs.

All of this has enormous implications for our interactions with nature, namely that humans cannot arrogantly or blindly force nature into our mold.[18] To cooperate with nature in a Taoist manner requires a better understanding of and appreciation for nature's processes. While an extreme Taoist position might advocate complete noninterference with nature, a more moderate Taoist approach would call for interaction with nature in a far less exploitive manner. Such cooperation with nature would sanction the use of appropriate or intermediate technology when necessary and would favor the use of organic fertilizers and natural farming methods. In terms of economic policy it would foster limited growth within a steady state economy that could support sustainable not exploitive development. Clearly, a Taoist ecological position is one with significant potential in the contemporary world.

Confucianism and Ecology: Cosmology and Ethics

Let us turn to the early classical texts of the *Analects* and the *Mencius* to explore the ecological dimensions of Confucian thought.[19] These works have had an enormous impact on Chinese society, education, and government for over two millennia. Along with two shorter texts, the *Great Learning* and the *Doctrine of the Mean*, these became known as the Four Books and were the basis of the civil service examination system from the fourteenth century until the twentieth.

Cosmologically, early Confuciansm, like Taoism, understood the world to be part of a changing, dynamic, and unfolding universe.[20] The ongoing and unfolding process of nature was affirmed by the Confucians and seasonal harmony was highly valued. There is no common creation myth per se for the Confucians or Taoists.[21] Rather, the universe is seen as self-generating, guided by the unfolding of the Tao, a term the Confucians shared with the Taoists although with variations on its meaning in different contexts and periods. There is no personification of evil; instead, there is a balance of opposite forces in the concept of the yin and the yang.

Indeed, there is no radical split between transcendence and immanence such as occurs in the Western religions. In fact, it has become widely accepted that the sense of immanence rather than transcendence dominates both Taoist and Confucian thought. Although this needs qualification, it is true that the notion of "the secular as the sacred" was critical in Chinese philosophy and religion.[22] The significance of this view is that a balance of the natural and the human worlds was essential in both Taoist and Confucian thought. While the Taoists emphasized harmony with nature and downplayed human action, the Confucians stressed the importance of human action and the critical role of social and political institutions.

Within this cosmology certain ethical patterns emerged in Confuciansm which are distinct from Taoism. Examining these patterns may be helpful for our understanding of the ecological dimensions of Confucian thought. While Taoism can be characterized as a naturalistic ecology having certain affinities with contemporary deep ecology, Confucianism might be seen as a form of social ecology having some similarities with the contemporary movement of the same name.[23] Taoism is clearly nature centered, while Confucianism tends to be more human centered. Neither tradition, however, succumbs to the problem of egocentric anthropocentrism or radical individualism such as has been characteristic of certain movements in the modern West. Both have a profound sense of the importance of nature as primary. For the Taoists nature is the basis of nourishing individual life and for the Confucians it is indispensable for sustaining communal life.

A Confucian ethic might be described as a form of social ecology because a key component is relationality in the human order against the background of the natural order. A profound sense of the interconnectedness of the human with one another and with nature is central to Confucian thinking. The individual is never seen as an isolated entity but always as a person in relation to another and to the cosmos. A useful image for describing the Confucian ethical system is a series of concentric circles with the person in the center. In the circle closest to the individual is one's family, then one's teachers, one's friends, the government, and in the outer circle is the universe itself. In the Confucian system relationality extends from the individual in the family outward to the cosmos. This worldview has been described as an anthropocosmic one, embracing heaven, earth, and human as an interactive whole.[24] In Confucianism from the time of the early classical text of the *Book of History*, heaven and earth have been called the great parents who have provided life and sustenance.[25] Just as parents in the family deserve filial respect, so do heaven and earth.[26] Indeed, we are told they should not be exploited wantonly by humans.

In Confucianism, then, the individual is both supported by and supportive of those in the other circles which surround him or her. The exchange of mutual obligations and responsibilities between the individual, others in these circles, and the cosmos itself constitute the relational basis of Confucian societies. Like a social glue, the give and take of these relationships help to give shape and character to these societies. Many of these patterns of social and cosmological exchanges become embedded in rituals which constitute the means of expressing reciprocal relations between people and with nature. Thus the value of mutual reciprocity and of belonging to a series of groups is fostered in Confucian societies. In all of this, education was critical. As the *Great Learning (Ta Hsueh)* so clearly demonstrates, to establish peace under heaven we must begin with the cultivation of the mind-and-heart of the person.[27] Education for the Confucians embraced the moral and intellectual dimensions of a person and was intended to prepare them to be a fully contributing citizen to the larger society.

In addition to these ethical patterns of social ecology for the individual in relation to others and to the cosmos itself, Confucianism developed an elaborate theory of government which might be described as a political ecology. Taking the same model of the individual embedded in a series of concentric circles, the Confucians situated the emperor at the center and suggested that his moral example would have a rippling effect outward like a pebble dropped into a pond. The influence of this morality would be felt by all the people and humane government would be possible when the emperor had compassion on the people and established appropriate economic, social, and ecological policies.

Thus while both Confucianism and Taoism are relational in their overall orientation, Confucianism is clearly more activist, especially with regard to moral leadership and practical policies. Many of the principles of humane government such as those advocated by Mencius and other Confucians include policies such as an ecological sensitivity to land and other resources, equitable distribution of goods and services, fair taxation, and allowing the people to enjoy nature and cultivate human relations.[28] The recognition that humane government rests on sustainable agriculture and maintaining a balance with nature is key to all Confucian political thought.

Thus both in terms of individuals and society as a whole there was a concern for larger relationships that would lead toward harmony of people with one another and with nature, which supported them. This social and political ecology within an anthropocosmic worldview has something to offer in our own period of rampant individualism, self-interested government, and exploitation of natural resources. The continuation of cooperative group effort to achieve common goals that are for the benefit of the whole society is an important model for a new form of social ecology. At the same time the ideal of humane government which develops and distributes resources equitably is central to a political ecology so much needed at the present.

Conclusion

This essay only begins to suggest some of the rich resources available in the traditions of Taoism and Confucianism for formulating an ecological cosmology and an environmental ethics in our time. As we seek a new balance in human-earth relations, it is clear that the perspectives from other religious and philosophical traditions may be instrumental in formulating new ways of thinking and acting more appropriate to both the vast rhythms and the inevitable limitations of nature. As our worldview in relation to nature is more clearly defined, we can hope that our actions will reflect both a Taoist appreciation for natural ecology and a Confucian commitment to social and political ecology.

Notes

1. *Commentary on the Lao Tzu*, trans. Wing-tsit Chan, in *A Source Book in Chinese Philosophy* (Princeton: Princeton University Press, 1963), 321.

2. *Mencius* 1A:5, trans. Wing-tsit Chan, *A Source Book in Chinese Philosophy*, 61.

3. Thomas Berry wrote: "The religious traditions must not only become acclimated to the new scientific and technological environment, they must undergo a breathtaking expansion of their own horizons as they become universalized, mutually present to each other, and begin creation of the multiform global religious tradition of humankind, a tradition that is already further advanced in the realities of human history than in the books we write." From "Future Forms of Religious Experience," in Berry's Riverdale Papers (an unpublished collection of essays).

4. Tu Wei-ming, "Beyond the Enlightenment Mentality: An Exploration of Spiritual Resources in the Global Community," a paper presented at the Fourth Conference on World Spirituality, East-West Center, Honolulu, June 15–19, 1992, and adapted for publication in this volume of *Bucknell Review*.

5. The Chinese have not had a strong environmental record in the modern period as demonstrated by Vaclav Smil, *The Bad Earth* (Armonk, N.Y.: Sharpe, 1984) and Lester Ross, *Environmental Policy in China* (Bloomington: Indiana University Press, 1988).

6. Charlene Spretnak has attempted such an examination in her comprehensive book *States of Grace* (San Francisco: Harper San Francisco, 1991). Another significant contribution is J. Baird Callicott and Roger T. Ames, eds., *Nature in Asian Traditions of Thought* (Albany: State University of New York Press, 1989).

7. This was also one of the aims of the Parliament of World Religions held in Chicago in September 1993.

8. Karl Jaspers describes the Axial period in his book *The Origin and Goal of History* (New Haven: Yale University Press, 1953).

9. For a more detailed account of this see Frederick Mote, *Intellectual Foundations of China* (New York: Knopf, 1971), chap. 2, and Tu Wei-ming's essays in *Confucian Thought: Selfhood as Creative Transformation* (Albany: State University of New York Press, 1985), esp. "The Continuity of Being: Chinese Visions of Nature," 35–50.

10. Indeed, Just as Rene Dubos noted the need for both a "passive" Franciscan model of reverence for nature along with the more "active" model of Benedictine stewardship in the West, so too can Taoism and Confucianism provide this complementary model. See Rene Dubos, "Franciscan Conservation versus Benedictine Stewardship" in *A God Within* (New York: Scribner's, 1972).

11. There are many editions of this text in English, including the James Legge translation (first published in 1891 and reissued by Dover in 1962), the Witter Bynner translation

(published by Capricorn in 1944), the D. C. Lau translation (published by Penguin in 1963). See also Wing-tsit Chan's translation in *A Source Book in Chinese Philosophy*.

12. The *Tao te Ching*, chap. 1. Unless otherwise noted, the translations here are Wing-tsit Chan's; subsequent references will be cited by chapter number in the text.

13. The term "deep ecology" implies that the human is deeply embedded in nature and not set apart from it. This term was coined by Arne Naess and is developed by Bill Devall and George Sessions in their book *Deep Ecology* (Salt Lake City, Utah: Peregrine Smith, 1985). Also, see the article in this volume by George Sessions on deep ecology.

14. Witter Bynner translates the opening of chapter 29 of the *Tao te Ching*: "Those who would take over the earth / and shape it to their will / Never, I notice, succeed." (Chan and Lau translate "earth" as "empire.")

15. See the *Tao te Ching*, chaps. 18, 19, 20, and 38 for examples of this.

16. See the *Tao te Ching*, chaps. 29, 37, 43, 48, 63 and 64.

17. See the *Tao te Ching*, chaps. 6, 10, 20, 25, 28, 32, 52, 55, 59, 78.

18. The degree of our contemporary hubris toward nature is revealed in an editorial essay in *Time*, 17 June 1991: "Nature is our ward. It is not our master. It is to be respected and even cultivated. But it is man's world. And when man has to choose between his well-being and that of nature, nature will have to accommodate."

19. See Arthur Waley's translation of the *Analects* (New York: Vintage Books, 1938) and D. C. Lau's translation of *Mencius* (Harmondsworth: Penguin Books, 1970).

20. See Tu Wei-ming's essays in *Confucian Thought*, esp. "The Continuity of Being"; also see Mote, *Intellectual Foundations of China*, chap. 2.

21. See Mote, *Intellectual Foundations of China*, 17–18.

22. See Herbert Fingarette, *Confucius—The Secular as Sacred* (New York: Harper Torch-books, 1972).

23. Social ecology has been developed by such theorists as Murray Bookchin. See his book *The Ecology of Freedom: The Emergence and Dissolution of Hierarchy* (Palo Alto, Calif.: Chesire Books, 1982). However, in contrast to Bookchin's position, which chooses to ignore the spiritual dimensions of ecology in the interest of establishing social justice in the human order, Confucianism tends to see the individual as embedded in a spiritual universe infused with *ch'i* (matter-energy). Harmonizing with the *ch'i* in the universe is essential for humans in a Confucian framework.

24. See Tu Wei-ming's use of this term in *Centrality and Commonality* (Albany: State University of New York Press, 1989) and in *Confucian Thought*.

25. See Book of Chou, The Great Declaration, *The Chinese Classics*, vol. 3, *Book of History*, trans. James Legge (Oxford: Clarendon Press, 1865), 283.

26. See Kaibara Ekken's *Yamato Zokkun*, translated in my book *Moral and Spiritual Cultivation in Japanese Neo-Confucianism* (Albany: State University of New York Press, 1989) 54–55, 136–42.

27. See the *Great Learning*, trans. Wing-tsit Chan, *Source Book in Chinese Philosophy*, 84–94.

28. See examples of humane government in book 1 of *Mencius*.

Contemporary Ecological Perspectives

The Emerging Ecological Worldview

Ralph Metzner

California Institute of Integral Studies

THE global environmental crisis is serving as a catalyst for far-reaching reexaminations of fundamental values and assumptions in every area of human knowledge and inquiry. Therefore, both challenge and opportunity exist for all the disciplines to reformulate some of the fundamental questions and issues in each field. Geologian Thomas Berry has said that the time has come to "re-invent the human at the species level." I take this to mean that the existing cultural paradigms cannot deal adequately with the issues we are now facing and that we need to draw on the evolutionary wisdom of the human species in its interrelationships with all other species and ecosystems. The viability of the human and its mode of adaptation to the natural world is now called into question, and indeed we have brought conditions on the entire biospheric life-system to a dangerous impasse.

It is not necessary to belabor the well-known parameters of the ecological catastrophe we are facing, as these are well-documented in publications such as the annual *State of the World* reports issued by the Worldwatch Institute.[1] The issues and problems of environmental pollution and degradation have passed from the literature of the scientific communities into the mainstream media. Pollution does not respect national boundaries, lending momentum to international, even global, cooperation. Ecosystem destruction does not respect sociopolitical boundaries, leading to new calls for social and ecological accountability from all levels of government and the professions, including law, business, medicine, and education. The study of environmental degradation and of environmental restoration does not respect the paradigm boundaries of traditional disciplines. Ecology, because of its concern with the complex web of interdependent relationships in ecosystems, including the pervasive role and impact of the human, is the interdisciplinary "subversive science" par excellence.

A growing chorus of voices is pointing out that the fundamental

163

roots of the environmental disaster lie in the attitudes, values, perceptions, and basic worldview that we humans of the industrial-technological global society have come to hold. The worldview and associated attitudes and values of the industrial age have permitted and driven us to pursue exploitative, destructive, and wasteful applications of technology. I suggest that we are in the midst of a transition phase to an ecological age, characterized by an ecological worldview, the outlines of which are being articulated in the natural sciences, the social sciences, and in philosophy and religious thought. The "Transition" chart at the end of this essay summarizes the main features of the emerging ecological worldview, which can be contrasted with the currently dominant industrial worldview shaped by the scientific revolution of the sixteenth and seventeenth centuries and the industrial revolution of the eighteenth and nineteenth.

Before moving on to a discussion of the emerging ecological worldview, let me briefly mention three other analyses of the transition we are presently undergoing. Whereas many social thinkers believe that the crucial transition taking place now is from the *industrial* era to the *information* or *electronic* era, I believe that electronic information technology is only the latest, most abstract, expression of the mechanistic, technological mindset and does not represent a real shift in values, such as ecology and the environmental crisis demand. Other critics argue that we are moving out of the *modern* age of rationalism and positivism, which began in the eighteenth-century enlightenment period, into a *postmodern* age of deconstructionist relativism, where all theories and models of reality are accorded equal validity and denied "privileged access." In contrast to this view, I believe it is possible to do more than to critique the modern view, and to recognize some consistent features of the newly emerging worldview, as they are (often separately) pursued and formulated in the different disciplines. Finally, the third reading of the current transition is that of Berry, who looks at the evolutionary time scale and proposes that we are moving out of the *Cenozoic* (age of mammals), which began sixty-five million years ago, into the *ecozoic* era, in which humans take their rightful place as members of the integral interdependent community of all life.[2]

In the *natural sciences*, several new paradigm transitions can be discerned. The "mechanical philosophy" of Newton, Galileo, and Descartes, which began by devising quantitative, mechanical models of physical processes, developed in the course of three centuries into a *mechanomorphic* worldview, in which the universe is erroneously identified with the analogical models originally designed to explain it. This mechanistic worldview is giving way in many circles to an *organismic* view, which sees the universe as an evolving process, a "story" in Berry's terms. Instead of seeing life as biochemical machinery somehow derived from random molecular combinations, the new biology defines life as a self-generating *(autopoietic)*, genetically coded process adaptively coupled with the environment. The Earth, instead of an inert body of dead matter, is seen in the Gaia theory of James Lovelock and Lynn Margulis as a kind of superorganism, evolving in homeostatic reciprocal interaction between living organisms and the physicochemical environment.[3] Some scientists have criticized the Gaia theory for not giving a new mechanism and only changing the metaphor. Such criticism ignores the fact that "mechanism" is itself a metaphor. The currently accepted *mechanomorphic* worldview is often not recognized as a metaphor. The psychic fixation of scientific thinking on the machine metaphor is demonstrated in even so eminent an ecologist as Paul Ehrlich, who can write a textbook with the title "The Machinery of Nature."

Quantum physics with its uncertainty principle has challenged the old deterministic model of a predictable clockwork universe. Traditional concepts of linear causality and mechanical forces acting on material objects are being superseded by chaos theory, nonlinear dynamics, and dissipative structures. The notion of chaos, as the epitome of unpredictable disorder, has been transformed by new mathematical approaches that yield unexpected orderliness in complex dynamical systems. The atomistic, or "billiard-ball," conception of ultimate reality is giving way to a holistic view, in which reality is analyzed as a *holarchy* (nested hierarchy) of systems with complex multilevel interactions of phenomena at all levels, from subatomic wave/particles and atoms to galactic clusters and universe.

In *epistemology*, the older view of logical positivism (sense observations are the only meaningful statements) and operationalism (the meaning of variables is in the experimental operations) has given way to more open-ended approaches that recognize the possible validity of different perspectives (critical realism) and that take into account the fact that theories and models are mental constructions (constructivism). The reductive-analytic strategy of doing scientific research, which looks for explanations "from below," has in the conventional paradigm led to a reductionist ontology, in which all the sciences are supposedly ultimately reducible to the physics of elementary particles. In the postmodern philosophy of science, the reductionist orientation is complemented by integrative, systemic perspectives, including the possibility of causation "from above."

The emerging ecological worldview involves a very different perception of the *role of the human* in the scheme of things. For thousands of years, since the beginnings of the Neolithic domestication, the human has tended to assume a dominating and exploitative attitude toward nature. Judeo-Christian theology has taught that man was created in God's image, put on earth to "subdue" and "have dominion" over the plants and animals. Heroic individualism and patrilineal property control have been the dominant value systems of the human, whose self-appointed task has been the "conquest of nature." The anthropocentric attitude assumes nature is an unlimited repository of resources, to be exploited for man's benefit, or, in the conservation ethic, to be conserved or managed for man's future uses.

In contrast, the influence of ecological concepts of co-evolution and symbiosis has led to an awareness of the evolutionary importance of protecting ecosystem integrity and preserving the diversity of species. The philosophy of deep ecology teaches biocentric or ecocentric values, in which humans are seen as part of nature, not over or against it. Philosophers of the deep ecology orientation suggest we have the potential of extending our sense of identity (identification) to include animals, plants, biotic communities, ecosystems, the earth.[4] The destiny of humankind is seen, then, not in the domination and control of nature but in the special quality of human consciousness, its unique reflectivity and tool-making creativity. Living systems of all kinds are valued intrinsically, in and for themselves, not instrumentally, as resources to be exploited, managed, or conserved. The most radical deep ecologists are not even comfortable with the stewardship notion, since it still implies that humans somehow have superior ecological knowledge and are therefore entitled to take care of the earth.

In *relation to the land*, the Western industrial-technological worldview is fundamentally based on the notion of property and ownership. Land exists to be used and developed for farming, herding, building, etc. Since the beginning of the Kurgan invasions of Europe and the Mediterranean by nomadic pastoralist warrior tribes from Central Asia (about six thousand years ago), the competing tribes have been fighting for territories and the herds and slaves that went with them. Primordial cultures such as the Native American have had a very different relationship to the land—more akin to stewardship, with a profound respect for place and the sacredness of certain particular places of power. The American ecologist Aldo Leopold spoke of a "land ethic" that would require us to learn to "think like a mountain."[5] The bioregional movement advocates a return to an appreciation of the natural (e.g., watershed) boundaries of a given region, optimally with decentralized self-sufficiency, and the human task is then to "re-inhabit" the place, to really know it and dwell in it.

The value systems governing *human social relationships* are also changing under the impact of the global transition to an ecological worldview. Feminists and ecofeminists, including Riane Eisler, author of *The Chalice and the Blade,* have cogently argued that the domination of nature in the Indo-European cultures that we have inherited is inseparable from the domination of women, who were seen as closer to nature, but who were regarded as possessions, along with the children, the herds, and the slaves of conquered peoples (often of other races or colors). Partnership, or "gylany," is the term used by Eisler to indicate the new, balanced male-female relationship pattern.[6] Value divisions based on racial or ethnic differences will increasingly give way to a new planetary culture which respects and celebrates qualitative differences. The beginnings of this can already be seen in the worldwide "fusion" of diverse styles in fashion, music, cuisine, and life-style, facilitated by the global media networks. The social ecology position of Murray Bookchin argues that class-domination patterns must be corrected simultaneously with the domination of nature.[7]

The *religious and theological* aspects of the environmental crisis have been pointed out by Lynn White, Jr., Arnold Toynbee, and others: in the three great monotheistic religions, God (always masculine) is a transcendent creator-and-law-giver deity, and there is an inseparable gulf between this God and humans, whose only recourse is to obey the law and support the priesthood or Church. In the animistic religious view of primordial peoples, all of nature—animals, plants, mountains, forests, streams, landscapes—was animated by living intelligences (called "spirits"), with which both shamans and ordinary people could be in communication. The monotheistic religions altered this forever: nature, the world, was the creation of a remote transcendent deity and was inherently corrupt; tainted by original sin, it was dark, nonsacred, and finally demonic and frightening (which fits with the command to dominate and conquer). By destroying pagan animism and the shamanic traditions preserved in witchcraft, Christianity drastically severed itself from the roots of a regenerative spirituality grounded in the natural world. Protestantism, which, as Max

Weber pointed out, furthered the development of exploitative capitalism by focusing on the value of work in the material world, completed the desacralization of the natural world. In the modern atheistic, materialist worldview, there is no spiritual being anywhere, either in this life or after death, either in nature or above it—but control, use, and exploitation are still the norm.

The polytheistic, animistic religions which preceded Judaism and Christianity, although their environmental record is not above reproach, at least still had a conception of spirituality as immanent within nature. Either pantheism ("everything is divine") or panentheism ("the divine is in everything") was the theology of the original Europeans and of the Jewish and Christian mystics (such as Francis of Assisi, Hildegard von Bingen) as well. "Creation spirituality," a concept formulated by the theologian Matthew Fox, is contrasted with the fall-and-redemption theology of mainstream Christianity.[8]

In thinking about *education and research*, the pursuit of knowledge has come to mean the ever-narrower specialization of disciplines and an unbridgeable gap between the "two cultures" of science and the humanities. The mechanistic paradigm of classical physics, which has been adopted by the life sciences and the social sciences, assumes that its method attains to "objective" knowledge, to "facts" free of values. Beginning with Thomas Kuhn, historians and philosophers of science have long since established that the pursuit of scientific knowledge is anything but free of values or metaphysical assumptions. In actuality the underlying value systems presupposed by science are congruent with the domination and exploitation agenda of the patriarchal mindset: prediction and control are the stated objectives of research, the results of which are fed into technology for "man's benefit" (read: profit and capital accumulation) and "security" (read: militarism). In the newly emerging worldview, with ecology as the model discipline, education and the pursuit of knowledge will of necessity be multidisciplinary and integrative. Unconscious values and hidden agendas will need to be brought into the light of critical review. Global citizens of a unified world in catastrophic transition cannot afford to hang on to the fragmentary paradigms of European industrial culture.

In the *political arena*, the industrial-technological culture has crystalized around the nation-state. During the modern era, the concept of nation-state sovereignty and centralized authority emerged out of the monarchic, feudal, and ecclesiastical forms of the medieval period. Patriarchal power groups, organized to protect patrilineal property and ownership "rights," imposed a gradually increasing stranglehold of industrial and militaristic cultural uniformity on their subject populations. The propagandistic use of mass-psychological processes of scapegoating and enemy-making culminated in the fascist, genocidal, totalitarian holocausts that European "civilization" inflicted upon the world in the twentieth century. In departing from these suicidal and ecocidal patterns, the kinds of political forms that are emerging are various forms of federations and confederations, a decentralization of the nation-state into pluralistic societies of ethnic and national groupings, increased reliance on self-sufficient and self-maintaining bioregions, and a shift of values and priorities away from military to human and environmental concerns.

The prevailing *economic systems*, both capitalist and socialist, are based on the illusion that unlimited material progress can be achieved by further industrialization, without any accounting of the earth's capital, and without concern for pollution, waste disposal, or species preservation. Under the impact of an avalanche of feedback that humans are exceeding the carrying capacity (the "limits to growth") of the biosphere, while destroying habitats and causing the extinction of countless species of plants and animals whose existence is vital to the regenerative capacity of the biosphere, these assumptions and policies are being abandoned in favor of cooperative, community-based, steady-state, sustainable economies that recognize the prime and ultimate dependence of all human economic activity as well as all nonhuman life forms on the integrity of the biosphere and the local ecosystems.

Profit-driven *technologies* that pollute the global elemental energy cycles and generate catastrophic amounts of toxic and nonrecyclable wastes will have to be replaced by appropriate technologies, also called "soft energy paths" by Amory Lovins,[9] and a massive conversion of the entire industrial infrastructure to reusable and recyclable materials and products. Technology, instead of being used to feed a runaway cycle of exploitation and consumerism ("more and more for more and more"), will need to be redirected toward the protection and restoration of damaged ecosystems.

In *agriculture* in the industrialized nations, excessive reliance on chemical fertilizers and pesticides, combined with monoculture using artificially produced hybrids, has led to disastrous loss of topsoil, genetic erosion, and decreasing yields for increasing populations. The way out of this dilemma, as propounded by the organic farming movement and thinkers such as Wes Jackson and

Wendell Berry, is to return to traditional, small- and medium-scale farming methods that used crop rotation and biological methods of pest control and achieved thereby a truly sustainable agriculture.

In reflecting on the ecological worldview outlined here, it would appear that there is actually a remarkable degree of congruence and agreement, if not consensus, even among people working in quite different areas. The disastrous features of our present policies and practices seem to flow from a few widely shared basic assumptions and value systems. These assumptions and values have no inherent staying power: they are cultural, not biological, givens. The alternate attitudes and values now being advocated in many circles are unconventional but not unnatural. Indeed, they seem to resonate to the most ancient human longings for exuberant life, freedom to grow, the recognition of spirit, the appreciation of differences, the delight in creativity. The pathways into the ecological age have been and are being convincingly articulated by many pioneers. It remains for us to muster the personal and political will to walk these paths.

TRANSITION FROM THE INDUSTRIAL TO THE ECOLOGICAL AGE

	Industrial Age	Ecological Age
scientific paradigms:	mechanistic	organismic
	universe as machine	universe as process/story
	Earth as inert matter	Gaia: Earth as superorganism
	life as random chemistry	life as autopoiesis
	determinism	indeterminacy, probability
	linear causality	chaos: nonlinear dynamics
	atomism	holism/systems theory
epistemology:	logical positivism	critical realism
	operationalism	constructivism
	reductionism	reduction/integration
role of the human:	conquest of nature	living as part of nature
	domination over nature	co-evolution, symbiosis
	individual vs. world	extended sense of self
	superiority & arrogance	reflection & creativity
	resource management	ecological stewardship
values in relation to nature:	nature as resource	preserve biodiversity
	exploit or conserve	protect ecosystem integrity
	anthropocentric/humanist	biocentric/ecocentric
	nature has instrumental value	nature has intrinsic value

	Industrial Age	*Ecological Age*
relation to land:	land use: farming, herding competing for territory owning "real estate"	land ethic: think like mountain dwelling in place reinhabiting the bioregion
human/social values:	sexism, patriarchy racism, ethnocentrism hierarchies of class & caste	ecofeminism, partnership respect & value differences social ecology, egalitarianism
theology & religion:	nature as background nature as demonic/frightening transcendent divinity creation as fallen, corrupt monotheism & atheism	animism: everything lives nature as sacred immanent divinity creation spirituality pantheism & panentheism
education & research:	specialized disciplines "value-free knowledge" pursued science-humanities split	multidisciplinary, integrative unconscious values explicated unified worldview
political systems:	nation-state sovereignty centralized national authority patriarchal oligarchies cultural homogeneity national security focus militarism	multinational federations decentralized bioregions egalitarian democracies pluralistic societies humans & environment focus commitment to nonviolence
economic systems:	multinational corporations assume scarcity competition limitless progress "economic development" no accounting of nature	community-based economies assume interdependence cooperation limits to growth steady state, sustainability economics based on ecology
technology:	addiction to fossil fuels profit-driven technologies waste overload exploitation/consumerism	reliance on renewables appropriate technologies recycling, reusing protect & restore ecosystems
agriculture:	monoculture farming agribusiness, factory farms chemical fertilizers & pesticides vulnerable high-yield hybrids	poly & permaculture community & family farms biological pest control preserve genetic diversity

Notes

This article was previously published as "The Age of Ecology" in *Resurgence*, no. 149 (November/December 1991).

1. Lester Brown et al., *State of the World 1993: Worldwatch Institute Report on Progress Toward a Sustainable Society* (New York: Norton, 1993).

2. Brian Swimme and Thomas Berry, *The Universe Story: From the Primordial Flaring Forth to the Ecozoic Era—A Celebration of the Unfolding of the Cosmos* (San Francisco: Harper San Francisco, 1992).

3. James Lovelock, *The Ages of GAIA: A Biography of Our Living Earth* (New York: Norton, 1988).

4. Bill Devall and George Sessions, *Deep Ecology: Living as if Nature Mattered* (Salt Lake City, Utah: Peregrine Smith, 1985).

5. Aldo Leopold, *A Sand County Almanac and Sketches Here and There* (New York: Oxford University Press, 1949).

6. Riane Eisler, *The Chalice and the Blade: Our History, Our Future* (San Francisco: Harper San Francisco, 1987).

7. Murray Bookchin, *Remaking Society: Pathways to a Green Future* (Boston: South End Press, 1990).

8. *Illuminations of Hildegard of Bingen,* with commentary by Matthew Fox (Santa Fe, N.M.: Bear & Co., 1985).

9. Amory Lovins, *Soft Energy Paths: Towards a Durable Peace* (Harmondsworth: Penguin Books, 1977).

Cosmology and Ethics

Larry L. Rasmussen
Union Theological Seminary

If you put God outside and set him vis-à-vis his creation and if you have the idea that you are created in his image, you will logically and naturally see yourself as outside and against the things around you. And as you arrogate all mind to yourself, you will see the world around you as mindless and therefore not entitled to moral or ethical consideration. The environment will seem to be yours to exploit. Your survival unit will be you and your folks or conspecifics against the environment of other social units, other races and the brutes and vegetables.

If this is your estimate of your relation to Nature and *you have an advanced technology,* your likelihood of survival will be that of a snowball in hell. You will die either of the toxic by-products of your own hate, or, simply, of overpopulation and overgrazing. The raw materials of the world are finite. If I am right, the whole of our thinking about what we are and what other people are has got to be restructured.[1]

T HESE sentences are not those of a theologian, philosopher, or ecologist, but an anthropologist, Gregory Bateson. They were not written in 1992, but 1972. They speculate—and warn— that a particular religious worldview and a way of life in keeping with it—when armed with the powers of modern technology— will do us in. Bateson concludes that an alternative worldview must be put in place.

In the discussion to follow, the intiguing matter is Bateson's assumed connections. He assumes the reality and efficacy of a cosmology and its ethic ("the whole [way] of our thinking about what we are and what other people are"). And he assumes the ecocrisis is a foundational challenge to the reigning cosmology because its consequences expose this cosmology and ethic as a death-dealing one.

In my judgment, Bateson is right. Others concur, although they express it differently. "Nothing less than the current logic of world

173

civilization"[2] runs counter to the well-being of the earth, Vice President Albert Gore writes. This logic drives not only the massive achievements of the industrial and postindustrial world but the ecocrisis itself. In searching for the cause, Gore says that "the more deeply I search for the roots of the global environmental crisis, the more I am convinced that it is an outer manifestation of an inner crisis that is, for lack of a better word, spiritual" (EB, 12). The spiritual crisis rests in the alienated way in which we conceive ourselves apart from nature. "We have misunderstood who we are, how we related to our place within creation, why our very existence assigns us a duty of moral alertness to the consequences of what we do" (EB, 258). Gore ends his book with his own statement of Christian faith as the reason for the hope that is in him and as the set of ultimate beliefs which buoys up his own part in the collective action "to change the very foundation of our civilization" (EB, 14). Beyond his testimony to Christianity he goes on to say of faith in general that its "essence" is "to make a surrendering decision to invest belief in a spiritual reality larger than ourselves. . . . Faith is the primary force that enables us to choose meaning and direction and then hold to it despite all the buffeting chaos in life" (EB, 368). Not surprisingly, the following paragraph is about the moral posture of faith and the need to accept moral responsibility for all choices, large and small. "It's the same gyroscope," Gore says, and closes by citing Aristotle that "virtue is *one* thing" (368).

In brief, Gore seems to assume Harold W. Wood, Jr.'s claim that "insofar as ordinary people are concerned, it is religion which is the greatest factor in determining morality,"[3] and he seems to mean by "spiritual" what others mean by "worldview," "cosmology," and "ethics"—namely, "the collection of values and assumptions that determine our basic understanding of how we fit into the universe" (EB, 12). Here are Bateson's assumptions and argument in another, extended form.

As subjects of debate and appeal, cosmology, ethics, religion, faith, the spiritual, and moral responsibility turn up with increasing frequency as the ecocrisis deepens. Their down-to-earth role may be judged a largely negative one, as Bateson's opening citation implies, or as Lynn White, Jr.'s now-famous essay on Western Christianity contended: Christianity "bears a huge burden of guilt" for the ecocrisis.[4] Yet religious cosmologies are consistently called upon to muster a response commensurate with the expanding challenge.[5] It is as though ecocrisis consciousness and religious consciousness are, in our time, like vast rivers which, though they arose in different terrain, now converge with one another in common channels.

This has been so from the very first wave of consciousness about the jeopardy facing the global environment. The M.I.T. report, *The Limits to Growth*, was one of the early marks. It was published in 1972, the same year as Bateson's book and John Cobb's *Is It Too Late? A Theology of Ecology*. Jørgen Randers's conclusion when he finished work on *Limits to Growth* was: "Probably only religion has the moral force to bring about [the necessary] change."[6] Mihajilo Mesarovic and Eduard Pestal soon followed (in 1974) with the first of several Club of Rome studies, *Mankind at the Turning Point*. It argued that "drastic changes in the norm stratum . . . are necessary in order to solve energy, food, and other crises."[7] Neither Randers nor Mesarovic and Pestal considered themselves "musical" in either religion or ethics and did not, at the outset, anticipate the implications of their own computer-modeled studies. More recently (January 1990) thirty-four internationally renowned scientists led by Carl Sagan and Hans Bethe issued "An Open Letter to the Religious Community." After detailing horrific environmental deterioration ("what in religious language is sometimes called 'Crimes Against Creation'"), the letter goes on:

> Problems of such magnitude, and solutions demanding so broad a perspective must be recognized from the outset as having a religious as well as a scientific dimension. Mindful of our common responsibility, we scientists—many of us long engaged in combatting the environmental crisis—urgently appeal to the world religious community to commit, in word and deed, and as boldly as required, to preserve the environment of the Earth.[8]

Further down the page the scientists say what modern scientists have, as scientists, been reluctant to say: "Efforts to safeguard and cherish the environment need to be infused with a vision of the sacred."

The "Open Letter" sparked a response which took the form of the Joint Appeal in Religion and Science. From the Joint Appeal itself came "The Summit on Environment" in June 1991, a gathering of religious leaders and scientists. The following was included in the emphatic summit statement written by religious leaders:

> Much would tempt us to deny or push aside this global environment crisis and refuse even to consider the fundamental changes of human

behavior required to address it. *But we religious leaders accept a prophetic responsibility to make known the full dimensions of this challenge. . . .*

Furthermore, *we believe a consensus now exists, at the highest level of leadership across a significant spectrum of religious traditions, that the cause of environmental integrity and justice must occupy a position of utmost priority for people of faith.*[9]

Whether this consensus exists or not, the appeal to religion in effect to jumpstart a deep and sustained response to the challenge of the ecocrisis is now a common one. But why is this so? It is certainly not self-evident, especially in view of the fact that so many who now make their religious appeal public are not religiously observant themselves. They have in fact been the most outspoken among those who (rightly) indict religious worldviews and practices for aiding and abetting ecocatastrophe. Why are they so ready to try yet another time?

The crude but real reason we engage cosmologies and moralities is that we seem unable to do otherwise. We are, as a species, storytellers who refuse to stop short of the cosmic story itself, despite (or because of?) its pretentious dimensions. This habit, as ancient as the oldest human records, may be unique to Homo sapiens and apparently rests in the trait that our species is both self-aware and death-aware. We wonder about origins and worry about destiny. More to the point, we not only study nature and life processes, indeed cosmic processes, we also conceive and draw the imposing maps we call "cosmologies." We not only study nature and live in and from it, we also construct the very *ideas* of nature that make their way into the lexicon of different cultures in different epochs and locales. We not only ponder, we organize our ponderings into grand narratives that become part and parcel of a way of life we live. We not only pray, as primal utterance, but we think about prayer and the kind of communication it can be.

The cultural forms and content of our cosmologies can and do vary widely. But invariably they are saturated with religious symbols and myths and religious rites and reflection. A straightforward, exacting empirical cosmology is also possible, of course, and in our time this steady work of good science is highly sophisticated and very far-reaching, charting worlds too small for our sense experience and regions and realities too expansive for our imagination. But the ages-old and continuing forms remain tenaciously religious as well, for better and worse. Religious cosmologies cannot do what good science does, but they hold distinct advantages science does not. Because their language is mythic, they can tell of the end and purpose of existence without abstracting them from experience. Indeed, they speak mythically from and to our deepest experience. Furthermore, religious cosmologies characteristically trade in ultimates—ultimate origins, destiny, meaning, value. Science can venture close to these but cannot leap, as science, into the final interpretation of faith. Too, religious cosmologies tend to rest at the foundation of cultural structures as well as their pinnacle, and show their many faces in the most common forms of art and language as well as the most arcane and elite. Religious cosmologies also promise not only meaning but survival power, deliverance, healing, well-being. They offer not only the wisdom of tested experience but the deep knowledge of revelation. Not least, they endender hope, the very nerve of moral action itself. In sum, religious cosmologies are nothing if not thirsty for exhibiting, in symbols, and explaining, in ritual, rite and reflection, the totality of things!

Religious cosmologies and traditions, though highly varied and both continuous and discontinuous over time and across cultures, are always concrete. How we are set in the grand scheme of things and what we make of it is almost as diverse as life itself and is part of that diversity. Thus Charlene Spretnak can look at the ecocrisis as a manifestation of the modern and postmodern world and lament "the monstrous reduction of the fullness of *being* that the Earth community currently faces through the dynamics of an increasingly manipulative, globablized, consumption-oriented political economy based on rapacious growth,"[10] and at the same time turn with hope and enthusiasm to the gathered wisdom of varied traditions, all of which have cosmic riches to offer. The nature of the serene mind-self in Buddhism, Spretnak writes, cuts the cord of the continuous cravings of distraction, greed, and anxiety, and the perverse encroachments of fear, indifference, rage and ill will. It simultaneously seeks a way of profound unity in the communion of all things and in daily practices of nonharming action. Native American spirituality, like that of other indigenous peoples, knows everything in our life to be kin to us, everything the wondrous yield of a common cosmic birth and a common, sustaining earth. The universe is not a collection of objects to be used; it is a vast community of living subjects who, as part of nature itself, are each on intimate terms with the rest.[11] We live best—indeed we only live at all—when we hearken to the wisdom of nature's ways and let the mystery of the spheres envelop us.

Goddess spirituality is yet another religious cosmology and set of traditions and practices. It knows that the cosmic body, the sacred whole that is in and around us, is present and experienced in our own bodies. So we rightly honor and embrace both the Earthbody and our personal bodies as sacred. The divine is not in some distant seat of power, in this view, but in life immediately at hand. To live aright is to "come to our senses." Islam, Judaism, and Christianity, like these other traditions, are internally very diverse and ongoingly creative. Yet for "the peoples of the Book" there is a bold line from the beginning, a certain concentration, a focus on community and social justice as a God-given vocation itself. A way of life is to be lived which emphasizes righteousness and moral responsibility. This is responsibility for the stewardship of all life and especially the "other" as neighbor.[12] We are all children of God who share in the common good and contribute to it in response to the gracious gift of life from God.

Yet whatever the different threads in different traditions and cultures, religious cosmologies all immodestly take on all of life, bundled together. Moreover, they detail the manner of life that accords with these grand explanations and their promises of healing and fulfillment. They offer a "way," a "way of life." It is no coincidence at all, then, that "ritual" is from *rita*, the Sanskrit signifying "order" and the etymological source of both "rite" and "right." Telling the arching cosmic story, learning the great narrative and giving it ritual expression, is the "rite" which offers the "right" ordering of existence and the guidance for living the "right" way. Ethics and cosmology are inextricable, indissoluble.

Differently said, we simply will not allow our experiences to dissipate in purely private sentiments. We insist that the larger meanings of life and their implications for behavior find a public and institutional form, become the material of conscious socialization and precious inheritance, indeed make their way to God in prayer, song, celebration, and mourning. Yet the point is not only what cosmologies "do" and their intimacy with moralities and ritual. The point is that we human creatures would not be the species we are apart from these constructs. We are incorrigibly cosmic storytellers and without cosmologies we *literally would not know what to do.* It is no surprise at all, then, that when the reigning cosmology and morality fails us or threatens to do us and others in, the reply is not, "Be done, then, with all such supercilious pastimes!" Instead, the response is to grope for a different or a transformed grand narrative by which "the whole of our thinking about what we are and what other people are [might] be restructured." Thus, when the challenge is as far-reaching as the ecocrisis, commensurate with our life systems' own borders, we can be certain cosmologies and moralities will be invoked, drawn upon, reviewed, scrutinized, changed, and employed anew. Richard Niebuhr is right; we simply cannot do otherwise:

> In the critical moments we do ask about the ultimate causes and the ultimate judges and are led to see that our life in response to action upon us, our life in anticipation of response to our reactions, takes place within a society whose boundaries cannot be drawn in space, or time, or extent of interaction, short of a whole in which we live and move and have our being.
>
> The responsible self is driven as it were by the movement of the social process to respond and be accountable in nothing less than a universal community.[13]

Notes

1. Gregory Bateson, *Steps to an Ecology of Mind* (New York: Random House, 1972), 472. As a quotation, I retain the original, though the pronouns for God are male only in this case. My usual practice is to avoid sexist language when direct citation is not used. I am grateful to Anne Primavesi for the citation from Bateson, together with her discussion. See her *From Apocalypse to Genesis: Ecology, Feminism and Christianity* (Minneapolis: Fortress Press, 1991), 24–43.

2. Albert Gore, Jr., *Earth in the Balance: Ecology and the Human Spirit* (New York: Houghton Mifflin, 1992), 269; hereafter, *EB*, with page references cited in the text.

3. Harold W. Wood, Jr., "Modern Pantheism as an Approach to Environmental Ethics," *Environmental Ethics* 7 (Summer 1985): 151.

4. See Lynn White, Jr., "The Historical Roots of Our Ecologic Crisis," *Science*, no. 155 (1967), 1203–7.

5. See, for example, Robin W. Lovin and Frank E. Reynolds, *Cosmogony and Ethical Order* (Chicago: University of Chicago Press, 1985).

6. Jørgen Randers, "Ethical Limitations and Human Responsibility," in *To Create a Different Future: Religous Hope and Technological Planning*, ed. Kenneth Vaux (New York: Friendship Press, 1972), 32.

7. Mihajilo Mesarovic and Eduard Pestal, *Mankind at the Turning Point* (New York: Dutton, 1974), 54.

8. From: "An Open Letter to the Religious Community," 3. The letter is available from the Office of the Joint Appeal in Religion and Science, 1047 Amsterdam Ave., New York, N.Y. 10025.

9. From "The Summit on Environment" statement, also available from the Office of the Joint Appeal in Religion and Science.

10. Charlene Spretnak, *States of Grace: The Recovery of Meaning in the Postmodern Age* (San Francisco: Harper San Francisco, 1991), 9.

11. Thomas Berry frequently uses this language to describe the universe. See *The Dream of the Earth* (San Francisco: Sierra Club Books, 1988) and Brian Swimme and Thomas Berry, *The Universe Story* (New York: Harper San Francisco, 1992).

12. Spretnak, *States of Grace,* passim.

13. H. Richard Niebuhr, *The Responsible Self* (New York: Harper & Row, 1963), 88.

Critical and Constructive Contributions of Ecofeminism

Charlene Spretnak

T HE earth-body and the womb-body run on cosmological time. Just as the flow of earth's life-giving waters follows lunar rhythms, so too follow the tides of a woman's womb. No culture has failed to notice these connections or the related feats of elemental power: that the female can grow both sexes from her very flesh and transform food into milk for them, and that the earth cyclically produces vast bounty and intricate dynamics of the biosphere that allow life. Cultural responses to the physical connections between nature and the female range from respect and honor to fear, resentment, and denigration. Whatever the response, it is elaborately constructed over time and plays a primal, informing role in the evolution of a society's worldview.

The central insight of ecofeminism is that a historical, symbolic, and political relationship exists between the denigration of nature and the female in Western cultures. The field has grown immensely since the term (as éco-féminisme) was coined in 1972 by Françoise d'Eaubonne in La féminisme ou la mort. Women have come into ecofeminism in the United States from several directions, including the environmental movement, various types of alternative politics, and the feminist spirituality movement.[1] In recent years, a number of ecofeminist anthologies, as well as hundreds of articles, have been published.[2] This introductory article will present three main aspects of ecofeminism: philosophy, political activism, and spirituality.

Historical Background

With regard to European cultures, considerable archaeological evidence indicates that both the earth and the female were held in high regard in the Neolithic settlements prior to the Bronze

Age.[3] Ritual figurines of a stylized sacred female with incised patterns of water or with the head of a bird, for instance, reflect perceptions of inherent interconnectedness with nature and seemingly "obvious" honoring of the elemental power of the female. After 4500 B.C. the archaeological record reveals a radical shift. Graves were no longer roughly egalitarian between the sexes (with women having somewhat more burial items than men) but suddenly followed the barrow model of burial, wherein a chieftan is surrounded by the bodies of men, women, children, animals, and objects that he owned or controlled. The westward migrations of nomadic Indo-European tribes from the Eurasian steppes imposed in old Europe a warrior cult, the addition of fortifications around settlements, a patriarchal social system, and the transferral of the sense of the sacred from nature and the female to a distant sky-god, although not all societies followed this pattern, of course.[4]

From the Bronze Age onward, the denigration of nature and the female in European societies fluctuated but never disappeared. The Pythagoreans codified their influential table of opposites in which the female is linked with the negative attributes of formlessness, the indeterminate, the irregular, the unlimited—that is, dumb matter, as opposed to the (male) principles of fixed form and distinct boundaries. Aristotle considered females to be passive deformities. The intellectual prowess of the male, he felt, could reveal and categorize all forms and functions of organisms in nature. Later, the medieval cosmology ranked men above women, animals, and the rest of nature, all of which were considered to be entangled with matter in ways that the male spirit and intellect were not. The advent of modernity—created by the succession of Renaissance humanism, the Scientific Revolution, and the Enlightenment—shattered the holism (but not the hierarchical assumptions) of the medieval synthesis by framing the story of the human apart from the larger unfolding story of the earth community.[5] The "new mechanical philosophy" of the Scientific Revolution in the sixteenth and seventeenth centuries perceived the natural world as a clockwork that could be fully apprehended and mastered by (male) human intellect. The practitioners of empirical science used metaphors that express heady delight in assaulting nature in order to learn "her secrets." Ecofeminists and others have noted that similar metaphors and attitudes were used in the "trials" (legalistic rituals of patriarchal hysteria) that preceded witch burnings and other torture during the era of the new rationalism.[6]

Dualistic Thinking in Western Philosophy and Culture

The dualistic thinking that has shaped so much of the Eurocentric worldview is perhaps the central concern of ecofeminist philosophical and political analysis. Countless ramifications follow from the Eurocentric notion of "the masculine" being associated with rationality, spirit, culture, autonomy, assertiveness, and the public sphere, while "the feminine" is associated with emotion, body, nature, connectedness, receptivity, and the private sphere. The reductionism of this orientation is accompanied by several assumptions that are essential to patriarchy: that the cluster of attributes associated with the masculine is superior to that associated with the feminine; that the latter exists in service to the former; that the relationship between the two is inherently agonistic; and that a logic of domination over nature and the female should prevail among (male) humans in the "superior" configuration. The Eurocentric construction of masculinity hence is a reactive and unstable posturing to appear "not-nature" and "not-female." The patriarchal core of the Eurocentric worldview is the culturally imposed fear that nature and the elemental power of the female are potentially chaotic and engulfing unless contained by the will of the cultural fathers.

Ecofeminists feel that the above analysis is relevant to identifying problematic assumptions in philosophical and political situations that have evolved within the Eurocentric orientation. In the area of ecofeminist philosophy, two topics that have received a good deal of attention are the critique of "manstream"[7] environmental ethics and the dialogue between ecofeminism and deep ecology.

Ecofeminist Critiques of Environmental Philosophy

With regard to the field of environmental ethics, ecofeminists maintain that many of its leading philosophers are largely blind to their patriarchal assumptions and hence can only replicate the logic of domination, albeit embedded in various versions of an ecological worldview. Ecofeminist philosophers reject the assumption that clinging to the rationalist concept of the self and the instrumental view of nature that dominates Western philosophy is a viable way to frame a postpatriarchal environmental ethics. The Kantian-rationalist framework is based on oppositionally construed reason: intellectual facility that is sharply distinct from the "corrupting" influences of the emotional, the personal, the particular.[8] Because the self is believed to be discontinuous from other humans and the rest of the natural world, moral progress is possible via a progression away from personal feelings to abstract, universalized reason. This approach results in strong opposition between care and concern for particular others (the "feminine," private realm) and generalized moral concern (the "masculine," public realm). Ecofeminists have identified this false opposition as a major cause of Western maltreatment of nature, noting that concern for nature should not be viewed as the completion of a process of (masculine) universalization, moral abstraction, and disconnection, discarding the self, emotions, and special ties ("Nature," passim).

Ecofeminists also challenge the Eurocentric concept of *rights* as a basis for philosophical frameworks of environmental ethics. "Ethical humanists" and "animal liberationists" attempt to establish the relative values of various parts of nature via such criteria as sentience, consciousness, rationality, self-determination, and interests. A being possessing one of these characteristics is said to have "intrinsic value" and hence the right to "moral consideration."[9] Ecofeminists generally regard this approach as static, arbitrary, and lacking a holistic apprehension of the natural world (including humans). Another objection to the use of rights theory is that it requires strong separation of individual rights-holders and is set in a framework of human community and legality. Its extension to the rest of the natural world often draws upon Mill's notion that if a being has a right to something not only should he or she (or it) have that something but others are obligated to intervene and secure it. Such reasoning gives humans almost limitless obligations to intervene massively in all sorts of far-reaching and conflicting ways in natural, balanced cycles to secure rights of a bewildering variety of beings ("Nature," 8). Ecofeminists feel that a more promising approach for an ethics of nature would be to remove the concept of rights from the central position it currently holds and focus instead on less dualistic moral con-

cepts such as respect, sympathy, care, concern, compassion, gratitude, friendship, and responsibility ("Nature," 9).[10]

The Ecofeminist Dialogue with Deep Ecology

The response of ecofeminist philosophers to the body of
thought known as deep ecology has drawn attention to its gender-
blind assumptions in condemning anthropocentrism without taking seriously the formative dynamics of androcentrism, or male
dominance. Most ecofeminists acknowledge common ground with
deep ecology's rejection of rationalist value theories and an environmental ethic grounded in abstract principles and universal
rules believed to be discoverable through reason alone.[11] Most
ecofeminists also appreciate deep ecology's rejection of the Eurocentric sense of discontinuity between humans and nature. However, ecofeminists are wary of assumptions that may lie embedded
in the concept of the "ecological self," which was formulated by
the founder of deep ecology, Arne Naess, and which refers to the
aspect of one's being that is continuous with the large Self (that
is, the unitive dimension of being) rather than the individual self.
It is sometimes described by Naess's colleagues in ways that could
be interpreted to result in the obliterating of all particularity, a
worrisome notion to the sex that has been socialized in patriarchal
culture to sacrifice their own self-definition to the needs of hus-
bands and children. Other ecofeminist concerns include issues of
differentiation (embedded in relationship), biocentric egalitarianism (the recognition that all species have worth), and concepts
of caring.[12]

Political Analysis and Activism

In this area, ecofeminists have astutely critiqued the masculinist
bias in the daily functioning of the environmental movement;[13]
played an important role in the growing challenge to the modern
model of "development" for the Third and Fourth Worlds;[14] and
been a leading force in campaigns for animal rights[15] and opposition to several aspects of reproductive technologies.[16] Most ecofeminist activists are engaged with grassroots political work,
whether or not they identify themselves with any particular party,

movement, or ideology. Many ecofeminists work in the Green politics movement, often within Green parties, because the democratic, community-based, and the ecological Green political vision
includes ecofeminist concerns and aspirations.[17] Its ideal of community-based economics, in which wealth and ownership are
spread as broadly as possible, stands in stark contrast to the increasing centralization of power and control in the hands of huge
corporations.

The experience of most feminists who have entered the environmental movement—either in institutionalized organizations or
alternative groups—has been painfully disillusioning. The historical link noted by ecofeminist theory between patriarchal attitudes
and the logic of domination over nature, women, and people of
color has yet to be acknowledged in practice by most male activists.
In the Green politics movement this situation is often somewhat
better than in environmentalist organizations because feminist
values are among the core values, at least in principle. Hence
Greens are ideologically committed to eliminating patriarchal behavior. When that fails to occur, women leave the Greens, as they
have demonstrated in several countries. Sometimes they return
when conditions improve.

An example of the political issues addressed by ecofeminists is
their vocal opposition to policies that reduce women of the Third
("developing") and Fourth (indigenous) Worlds to "resources" in
the emerging global economy. A leading ecofeminist critic, Vandana Shiva, who is an Indian physicist, maintains that "maldevelopment" is a new project of Western patriarchy, one that results in
the death of the "feminine principle." She asserts that the modern
model of development being imposed by the West is inherently
patriarchal because it is fragmented, "anti-life," opposed to diversity, dominating, and delights in "progress" based on nature's destruction and women's subjugation.[18] Ecofeminists insist that
Third and Fourth World women themselves must have control
over decisions about whether to opt for local self-reliance or integrate into the global economy. Unfortunately, enormously powerful banks and transnational corporations in the "developed" world
are furthering "maldevelopment" via centralized, large-scale projects that are usually capital-intensive, energy-intensive, and dis-
ruptive of local self-reliance and ecological integrity.

Spiritual Dimension

In addition to the philosophical and political aspects, ecofeminism contains a spiritual dimension. The ecofeminist alternative to the Western patriarchal worldview of fragmentation, alienation, agonistic dualisms, and exploitative dynamics is a radical reconceptualization that honors holistic integration: interrelatedness, transformation, embodiment, caring, and love. Such an orientation is simpatico with the teachings of several Eastern and indigenous spiritual traditions on nonduality and the relative nature of seemingly sharp divisions and separations. To refer to the ultimate mystery of creativity in the cosmos—its self-organizing, self-regulating dynamics—spiritual traditions draw on metaphor and symbol. Those may be female, male, or nonanthropomorphic, such as the Taoist perception of The Way. Ecofeminists are situated in all the major religious traditions, and most see good reason for women to use female imagery in references to "the divine," or ultimate mystery in the cosmos. Particularly in patriarchal societies, the choice of female metaphors is a healthy antidote to the cultural denigration of women. Those ecofeminists drawn to Goddess spirituality appreciate the nature-based sense of the sacred as immanent in the earth, our bodies, and the entire cosmic community—rather than being located in some distant father-god far removed from "entanglement" with matter. The transcendent nature of creativity in the cosmos, or the divine, lies not above us but in the infinite complexity of the sacred whole that continues to unfold. Goddess spirituality is not the sole tradition that contains these understandings, and even religions that are somewhat hostile to them are being persistently challenged by their own ecofeminist members.

Conclusion

In summary, ecofeminism is a movement that focuses attention on the historical linkage between denigration of nature and the female. It seeks to shed light on *why* Eurocentric societies, as well as those in their global sphere of influence, are now enmeshed in environmental crises and economic systems that require continuing the ecocide and the dynamics of exploitation. Ecofeminism continues the progression within traditional feminism from attention to sexism to attention to all systems of human oppression (such as racism, classism, ageism, and heterosexism) to recognition that "naturism" (the exploitation of nature) is also a result of the logic of domination.[19] Ecofeminism challenges environmental philosophy to abandon postures upholding supposedly gender-free abstract individualism and "rights" fixations and to realize that human relationships (between self and the rest of the world) are constitutive, not peripheral. Hence care for relationships and contextual embeddedness provides grounds for ethical behavior and moral theory. Politically, ecofeminists work in a broad range of efforts to halt destructive policies and practices and to create alternatives rooted in community-based legitimacy that honors the self-determination of women as well as men and that locates the well-being of human societies within the well-being of the entire earth community. Spiritually, ecofeminists are drawn to practices and orientations that nurture experiences of nonduality and loving reverence for the sacred whole that is the cosmos.

Ecofeminism is a global phenomenon that is bringing attention to the linked domination of women and nature in order that both aspects can be adequately understood. Ecofeminists seek to transform the social and political orders that promote human oppression embedded in ecocidal practices. The work consists of resistance, creativity, and hope.

Notes

1. See Charlene Spretnak, "Ecofeminism: Our Roots and Flowering," in *Reweaving the World: The Emergence of Ecofeminism,* ed. Irene Diamond and Gloria Feman Orenstein (San Francisco: Sierra Club Books, 1990).

2. The anthologies include the one in the previous citation plus *Healing the Wounds: The Promise of Ecofeminism,* ed. Judith Plant (Philadelphia: New Society Publishers, 1989); *Ecofeminism: Women, Animals, and Nature,* ed. Greta Gaard (Philadelphia: Temple University Press, 1993); *Ecofeminism and the Sacred,* ed. Carol J. Adams (New York: Crossroads Press, 1993); *Ecofeminist Philosophy,* ed. Karen J. Warren (New York: Routledge, forthcoming); and *Ecological Feminism,* ed. Karen J. Warren (Bloomington: Indiana University Press, forthcoming).

3. See Marija Gimbutas, *The Language of the Goddess: Unearthing the Hidden Symbols of Western Civilization* and *The Civilization of the Goddess: Neolithic Europe before the Patriarchy* (San Francisco: Harper San Francisco, 1989 and 1991 respectively). Gimbutas cites the work of numerous European archaeologists in addition to the discoveries from her own excavations.

4. For an account of other, living options, see Peggy Reeves, *Female Power and Male Dominance* (Cambridge: Cambridge University Press, 1981).

5. See Thomas Berry, *The Dream of the Earth* (San Francisco: Sierra Club Books, 1988).

6. See Carolyn Merchant, *The Death of Nature: Women, Ecology, and the Scientific Revolution* (San Francisco: Harper & Row, 1980). Also see Susan Griffin, *Woman and Nature: The Roaring inside Her* (New York: Harper & Row, 1978).

7. The term "manstream" is used by Janis Birkeland to refer to the male-dominant mainstream of Eurocentric societies. See "Ecofeminism: Linking Theory and Practice," in Gaard, ed., *Ecofeminism*.

8. See Val Plumwood, "Nature, Self, and Gender: Feminism, Environmental Philosophy, and the Critique of Rationalism," *Hypatia: A Journal of Feminist Philosophy* 6, no. 1 (Spring 1991): 1–7; hereafter, "Nature," with page references cited in the text.

9. See Marti Kheel, "The Liberation of Nature: A Circular Affair," *Environmental Ethics* 7, no. 2 (Summer 1985): 139.

10. This list is common in feminist models for ethics, but Plumwood is citing here from a book on Buddhism, Francis Cook's *Hua-Yen Buddhism: The Jewel New of Indra* (University Park: Pennsylvania State University Press, 1977). Plumwood's citation of an admirable approach to moral theory from a book on Buddhism reflects the common ground between many of the concerns of ecofeminism and those spiritual traditions that emphasize nonduality, wisdom, and compassion.

11. Marti Kheel, "Ecofeminism and Deep Ecology: Reflections on Identity and Difference," *Covenant for a New Creation: Ethics, Religion, and Public Policy*, ed. Carol S. Robb and Carl J. Casebolt (Mary Knoll, N.Y.: Orbis Books, 1991), 142–45.

12. For a summary of the dialogue between ecofeminism and deep ecology as of spring 1987, see Michael E. Zimmerman, "Deep Ecology and Ecofeminism: The Emerging Dialogue," in Diamond and Orenstein, eds., *Reweaving the World*. Also see the articles by Kheel and Plumwood cited in nn. 8, 9, and 11, above. Also see Charlene Spretnak, "Radical Nonduality in Ecofeminist Philosophy," in Warren, ed., *Ecological Feminism*.

13. See Pam Simmons, "The Challenge of Feminism," *The Ecologist* 22, no. 1 (January/February 1992): 2–3.

14. See Vandana Shiva, *Staying Alive: Women, Ecology and Development* (London: Zed Books, 1988). Also see Pam Simmons, "'Women in Development': A Threat to Liberation," *The Ecologist* 22, no. 1 (January/February 1992): 16–21.

15. See Carol J. Adams, *The Sexual Politics of Meat* (New York: Continuum, 1990). Also see Andree Collard and Joyce Contrucci, *Rape of the Wild: Men's Violence against Animals and the Earth* (London: The Women's Press, 1988). Also see several articles in Gaard, ed., *Ecofeminism*. Also see the article by Kheel cited in n. 9, above.

16. See Irene Diamond, "Babies, Heroic Experts, and a Poisoned Earth," *Reweaving the World*, 201–10. Also see Vandana Shiva, "The Seed and the Earth: Women, Ecology, and Biotechnology," *The Ecologist* 22, no. 1 (January/February 1992): 4–7.

17. The "Ten Key Values" of the Green politics movement in the U.S. are ecological wisdom, nonviolence, grassroots democracy, social justice and personal responsibility, community-based economics, decentralization, feminism, respect for diversity, global responsibility, and sustainable future focus. See Charlene Spretnak and Fritjof Capra, *Green Politics: The Global Promise* (New York: Dutton, 1984).

18. See Shiva, *Staying Alive*.

19. See Karen J. Warren, "The Power and Promise of Ecological Feminism," *Environmental Ethics* 12, no. 2 (Summer 1990): 132–46. Note that the first footnote in this article is a bibliographic listing that cites numerous ecofeminist books and articles.

Whitehead's Deeply Ecological Worldview

David Ray Griffin

School of Theology at Claremont and Claremont Graduate School

ALFRED North Whitehead (1861–1947) once described Christianity as "a religion seeking a metaphysic."[1] It can equally be said that environmentalism is a movement seeking a worldview. Valuable ideas for an appropriate worldview can, as this volume illustrates, be drawn from many philosophies, theologies, and traditions, including modern-becoming-postmodern science. After this is realized, however, the question still remains as to which overall worldview provides the best standpoint from which to appropriate elements from the others.

My own judgment is that Whitehead himself went far toward providing the kind of worldview that the environmental movement needs. The kind of worldview that is needed, I believe, is one that is deeply ecological; one that is pragmatic, in the sense of providing a livable guide for action; one that can be commended, because of its coherence and relative adequacy, as at least not obviously false; and one that, as part of the evidence for its relative adequacy, can reconcile tensions between other positions, doing justice to the elements of truth in each. Whitehead's cosmological philosophy has, I believe, all these virtues.

Before beginning the exposition of Whitehead's worldview, I will explain, in a preliminary way, how I am using the term "deeply ecological," and point out how Whitehead's position can reconcile a central conflict between some views that can be thus characterized.

Introduction

The term "deep ecology" is ambiguous, even embattled.[2] In its most general use, it refers to any environmental ethic that is not purely anthropocentric—that bases the call for environmental preservation and restoration not solely on enlightened human self-interest but also on the intrinsic value of other species.[3] We can call this deep ecology$_{na}$ (nonanthropocentric deep ecology). In this most general sense, those who focus on the rights of animals are deep ecologists. But that is not the way the term is usually employed: "animal liberationists" have usually been contrasted with "deep ecologists."[4] Not only do animal liberationists, as the name implies, typically focus only on individual animals, rather than on other forms of life and ecosystems, most of them also limit their concern primarily to the highest forms of animal life, especially fellow mammals; furthermore, most of them draw a line below which there is assumed to be no reason for ethical concern. Deep ecologists, by contrast, are typically concerned with the biosphere as a whole and do not draw a line beneath which there is assumed to be no inherent value to be respected. We can call this deep ecology$_b$ (biospheric deep ecology). For some deep ecologists, finally, even this stipulation is not sufficiently precise. To be a truly deep ecologist, say some followers of Arne Naess (who coined the term "deep ecology"), one must affirm "biospherical (or biological) egalitarianism," rejecting any type of hierarchy of value according to which some beings have more inherent value than others. We can call this deep ecology$_e$ (egalitarian deep ecology).[5]

In speaking of Whitehead's worldview as deeply ecological, I mean, in the first place, that his position supports deep ecology in the first two senses: deep ecology$_b$ as well as deep ecology$_{na}$. Nevertheless, while supporting biospheric deep ecologists over against any dualistic line-drawing and any exclusive focus on individuals in distinction from ecosystems, Whitehead's position also implies that the animal liberationist position, in presupposing that the higher animals are worthy of special concern, is rooted in a sound intuition. A synthesis of deep ecological$_b$ and animal liberationist positions is thereby achieved.

This synthesis is possible, however, only because Whitehead's philosophy rejects deep ecology$_e$ as hitherto usually understood, and has therefore been *contrasted* with deep ecology by both Whiteheadians and deep ecologists.[6] That is, Whitehead's position says that some types of beings have more capacity to realize intrinsic value than others, that this greater capacity includes both a greater range of potentialities that can be realized and a greater capacity for suffering, and that it is ethically appropriate, accord-

ingly, to be especially concerned about not needlessly causing suffering in such beings or preventing the realization of their desires.

This position does not mean, however, that Whitehead's
worldview is necessarily opposed to the intuition on which egalitarian deep ecology is based. I said above that it is opposed to
deep ecology$_e$ "as hitherto usually understood." Whitehead's intuition that there are different levels of intrinsic value may be compatible with the intuition of egalitarian deep ecologists that all
things, at least all living things, have equal inherent value. The
possibility of reconciliation arises from the fact that what the advocates of deep ecology have meant in speaking of *inherent* value
(whether they use the term "inherent" or "intrinsic") may differ
from what Whiteheadians mean in referring to *intrinsic* value.

Given this introduction, I shall, in the following sections, show
that Whitehead's worldview is deeply ecological in the following
senses: (1) it portrays all individuals as having intrinsic value; (2)
it portrays all things as internally related to their environments;
(3) it portrays the self in particular as an ecological self; (4) it
portrays the divine reality as ecologically interconnected with the
world and shows that the support given to an ecological consciousness by this portrayal of the divine reality is not undermined by
the problem of evil; and (5) it shows how a special concern for
human beings and other higher animals is not inconsistent with
concern for the biosphere as a whole and with the intuition that,
in some sense, all forms of life have (roughly) equal inherent value.

Intrinsic Value

Although the term "intrinsic value" was used only occasionally
by Whitehead himself, it has been widely used by Whiteheadians
as a technical term for the value that something has *in and for
itself.* The only things that are anything *for* themselves are, of
course, things with experience. The term "intrinsic value" stands
in contrast with "extrinsic value," which means any kind of value
something has for anything else. (The term "instrumental value"
has widely used in this inclusive sense, but that term is better
saved for only one of the many kinds of extrinsic value.) A most
important dimension of the extrinsic value of something is its
ecological value, meaning its value for sustaining the cycle of life.
Its value as food for other beings would be part of this ecological
value, but so would many other functions, such as the function of

worms in aerating the soil and that of certain soil bacteria in nitrogen fixation. Other forms of extrinsic value are *companion value, instrumental value* (in the narrow sense, such as a stick's value to a bird in ferreting out bugs from a tree limb), *aesthetic value,* and *medicinal value.* Some forms of extrinsic value are such only to human beings, such as *scientific value, monetary value,* and *symbolic* (including *moral* and *religious*) *value.* From these examples it can be seen that anything, whether or not it has intrinsic value (value for itself), can have extrinsic value (value for others).

The fact that only those things with experience can have intrinsic values does not, for Whitehead, mean that the world is divided into two kinds of actual things—those with and those without intrinsic value. Descartes and other dualists, of course, have assumed otherwise. For Descartes, the line (among earthly creatures) was to be drawn between human beings and everything else; other dualists draw the line further down, attributing experience to all animals with central nervous systems, or still lower. Whitehead believes, for various reasons, that no such line can be drawn.

One reason for affirming this view is theoretical: it is impossible to understand how experiencing things and nonexperiencing things could interact. The obvious example is the notorious mind-body problem bequeathed to modern philosophy by Descartes by his assumption that the brain is composed of matter devoid of experience. But even if one draws the line further down at, for example, the rise of life, the problem is equally severe: how can the experience of the cell influence, and be influenced by, its molecular parts, if those molecules be wholly devoid of experience?

The evolutionary picture of the world creates a new form of this theoretical problem: how could experience have evolved out of things wholly devoid of experience? It is often said that this is unproblematic, being simply one more example of "emergence": just as wetness emerges out of a combination of hydrogen and oxygen, neither of which is wet, so could experience emerge out of things that were completely devoid of experience. This argument, however, involves a category mistake. Wetness is a quality of things as they are *for others.* We do not suppose that the water molecules feel wet to themselves. "Experience," however, is what something is *for itself.* To say that experience arose out of a constellation of things without experience, therefore that things that exist *for themselves* arose out of things that were *nothing* for themselves, existing only *for* others, is to postulate an absolutely unique type of emergence, with no analogues.

This twofold theoretical difficulty with the dualistic hypothesis is buttressed by an empirical difficulty, which is that science has increasingly been removing the bases for assuming absolute discontinuity anywhere in the evolutionary process, whether with the rise of humans, or life, or anywhere else. In any case, Whitehead attributes experience to individuals all the way down. He does not mean, of course, that all types of individuals have the level of experience enjoyed by human beings, or even that enjoyed by dogs and cats. Even the kind of experience enjoyed by a living cell is already a fairly high-level experience, compared with that of macromolecules, ordinary molecules, atoms, and subatomic entities such as protons and electrons. The experience of such entities must be assumed to be very trivial. The point remains, however, that they should be thought to be only different in degree from us, however greatly, not wholly different in kind. The notion of "vacuous actuality"—the idea of something that is actual and yet "void of subjective experience"[7]—should be abolished.

Whitehead realized that this move had ethical implications. Kant had said that we should treat other human beings as ends in themselves, not merely as means to our own ends. The implication was that we *could* treat all other beings as simply means to our ends. "But if we discard the notion of vacuous existence," Whitehead points out, "we must conceive each actuality as attaining an end for itself."[8] In a chapter in which he expresses agreement with the Romantic poets, Whitehead says: "'Value' is the word I use for the intrinsic reality of an event."[9] The idea of vacuous actuality results from the "fallacy of misplaced concreteness," in which an abstraction is assumed to be the concrete reality (*SMW*, 51; *PR*, 7–8). "In physics there is abstraction. The science ignores what anything is in itself. Its entities are merely considered in respect to their extrinsic reality" (*SMW*, 153). Modern science, with its "Cartesian scientific doctrine of bits of matter, bare of intrinsic value" (*SMW*, 195), has had negative consequences, because it fostered "the habit of ignoring the intrinsic worth of the environment which must be allowed its weight in any consideration of final ends" (*SMW*, 196). To abolish the notion of vacuous actuality would be to establish the basis for replacing this bad habit with the habit of reverence:

Everything has some value for itself . . . . By reason of this character, constituting reality, the conception of morals arises. We have no right to deface the value experience which is the very essence of the universe.[10]

Given its denial of actualities devoid of experience, Whitehead's worldview can be called "panexperientialism" (although Whitehead himself used neither this term nor the more common "panpsychism").[11] The "pan," however, must not be understood to mean literally everything to which one can refer. The meaning is that all *actual individuals* experience. Excluded are things with merely *ideal* (rather than actual) existence, such as numbers, concepts, propositions, and *aggregates* of individuals that are not themselves individuals. This latter point cannot be too strongly stressed. Panexperientialist doctrines are regularly dismissed by pundits who suppose that they are saying something devastating by intoning: "rocks do not have feelings." But panexperientialism, at least of the Whiteheadian type, does not hold that they do.

What this cosmology does hold is that there are two basic ways in which individuals can be organized. On the one hand, individuals at one level can be organized so as to give rise to a higher-level individual that turns the whole into a "compound individual" (a term coined by Charles Hartshorne, the second major developer of this worldview).[12] Humans and squirrels are obvious examples, as the trillions of cells constituting the body give rise to a much higher-level individual, which we call the mind or soul, which turns this aggregation of cells into an individual. Cells, organelles, macromolecules, ordinary molecules, and atoms can also be considered compound individuals. On the other hand, an aggregation of individuals may not give rise to a higher-level experience, in which case the aggregate remains a mere aggregate. Sticks, stones, chairs, and mountains provide obvious examples. The only experiences in the stone are the experiences of the various molecules making it up; the stone qua stone (in contrast with the living squirrel qua squirrel) has no experience.

Making this distinction acknowledges the obvious distinction on which Descartes's dualism was based, yet does so without falling prey to the problems created by that dualism. The obvious distinction is between those objects of our sensory experience that do and those that do not seem to be capable of initiating activity and making a unified response to their environments—the distinction between sticks, stones, and shingles, on the one hand, and squirrels, snakes, and seagulls, on the other. We instinctively, and reasonably, assume that the latter, like ourselves, have a unified experience, while the former do not. Descartes's dualism is based on the assumption that the ultimate units of the world are tiny things that are more like stones than like squirrels. Whitehead avoids this dualism by making the opposite assumption, that mole-

196

cules are more analogous to squirrels and humans than to stones. Through this analogy, what we call nature is viewed as permeated by value: "We have only to transfer to the very texture of realisation in itself that which we recognise so readily in terms of human life" (*SMW*, 93).

Internal Relations

Besides intrinsic value all the way down, the other central point of any deeply ecological philosophy is a strong doctrine of internal relations, meaning relations that are internal to, and thus constitutive of, the things in question. Modern thought has been as unecological in this regard as the former. It has seen the world as comprised of substances that, in Descartes's notorious phrase, "need nothing but themselves in order to exist." Accordingly, Whitehead complains, in modern philosophies "the relations between individual substances constitute metaphysical nuisances: there is no place for them" (*PR*, 137). In Whitehead's philosophy, by contrast, relations are fundamental. Each thing arises out of its social relations and is internally constituted by these social relations. To see how fundamental social relations are in Whitehead's worldview, we need a more precise notion of his view of individuals.

I have thus far written of *enduring* individuals, such as electrons, molecules, cells, and squirrels. For Whitehead, however, the most fully actual things are not enduring individuals but momentary events. He calls them "actual occasions" or "occasions of experience." These occasions have a more-or-less brief existence, lasting anywhere from less than a billionth of a second, in the case of electronic or photonic occasions, to perhaps a tenth of a second, in the case of an occasion of human experience.

This doctrine means that things that endure, such as electrons, molecules, and minds, already exemplify a type of social existence. Whitehead, accordingly, calls them "societies":

> The real actual things that endure are all societies. They are not actual occasions. It is a mistake . . . to confuse societies with the completely real things which are the actual occasions.[13]

Enduring individuals are "temporally ordered societies": the social relations are purely temporal, because only one member exists

at a time. Any given occasion of experience belongs to that temporal society that consists of all those members that came before it
and all those that will come after it.

This doctrine makes social relations fundamental, while making
"enduring substances" derivative. What appears to be an independent substance, such as a proton, is in reality a pattern of
social relations, with perhaps a billion such relations occurring in
each second. Each actual occasion, however, does not simply arise
out of its predecessor in the temporally ordered society to which
it belongs. Each occasion is also influenced, even if less significantly, by other past occasions. In fact, each occasion is influenced,
to some slight degree, by the whole past universe. "The whole
world conspires to produce a new creation" (RM, 109).

In being affected by previous events, furthermore, an occasion
of experience is not simply affected in some external way, as billiard-ball images of causation suggest. Rather, the causal influence
of the past upon the present is in-fluence, a real in-flowing, which
affects the present experience internally. We know that our own
experience is influenced in this way. The causal influence of my
parents on me, for example, is nothing like the impact of two
billard balls upon a third. Rather, I am who I am in large part
because of attitudes, values, and habits that I internalized from
them. By thinking of all actualities as series of occasions of experience, we can think of all causal influence between individuals in
those terms.

"Prehension" is Whitehead's term for this internal appropriation of causal influences from the past. He says: "I use the term
'prehension' for the general way in which the occasion of experience can include, as part of its own essence, any other entity" (AI,
234). "The essence of an actual entity," he also says, "consists solely
in the fact that it is a prehending thing" (PR, 41). That is, far
from requiring nothing else to be, it is the very nature of an actual
entity to prehend previous actual entities into itself. This idea
that occasions of experience are taken up into later occasions of
experience is at the very heart of Whitehead's cosmology. His
philosophy, he says, "is mainly devoted to the task of making clear
the notion of 'being present in another entity'" (PR, 50).

This notion brings us to another distinctive feature of
Whitehead's cosmology, the idea that each actual entity exists in
two modes. It exists first as a subject of experience. In this mode,
it prehends prior experiences, then makes a self-determining response to them. (This element of self-determination may be more

or less significant.) After existing as a subject, its subjectivity perishes; it then exists as an object for subsequent subjects.

This twofold mode of existence is correlated with the previous distinction between intrinsic and extrinsic value. In referring to the intrinsic value of something, we are referring, strictly speaking, to the momentary actual occasions, not the enduring individuals. While it is a subject enjoying experience, an actual occasion has intrinsic value, value for itself. When it becomes an object for others, it has extrinsic value, value for others. It cannot have both kinds of value at the same time: it does not become an object for others until its moment of subjectivity is over. Only individual occasions can enjoy intrinsic value. But anything that can be prehended—an individual experience, a temporally ordered society of occasions, or an aggregation of occasions—can have extrinsic value.

The Ecological Self

Given the fact that human experience is taken as paradigmatic for actual entities, some aspects of the ecological nature of the self have been established by what has been said above: the enduring self is a temporally ordered society of occasions of experience, each of which arises out of its relations to former occasions of experience and then contributes itself to later occasions. Each moment of experience is a microcosm, taking into itself, at least to some slight degree, all prior events. For the momentary self to realize its true nature is to realize that it is akin to all other things. What remains here is to explain, if this *is* the true nature of the self, why this is not more obvious.

One reason for the perceived opposition between self and world has already been explained: low-grade actual occasions can be organized into aggregations in which evidence of their spontaneity, and thus of their experience, is masked. This ontological basis for the illusion that we are different in kind from "nature" is supported by an epistemological basis: in sensory perception, we do not perceive individual actualities, but only blurred masses. The things that we see and touch, accordingly, seem to be passive, hence entirely different from what we know ourselves to be—individuals with spontaneous energy (*AI*, 213, 218). The illusion that the "physical world" is entirely different from us is also promoted by the fact that sensory perception is a derivative form of perception. Sensory perception is so prominent in consciousness not, as modern philosophy has assumed, because it is our fundamental form of perception, but precisely because it is not. This is so because consciousness is itself a derivative form of experience.

Each occasion of experience consists of several phases. Consciousness is a special way of experiencing objects that may or may not be evoked in a late phase. Because consciousness arises, if at all, only in a late phase, it *tends to illuminate only those elements of experience that themselves arise in a late phase.* The negative point here is that consciousness therefore does *not* cast a bright searchlight upon those elements of experience that are truly fundamental in the sense of arising in the initial phase of experience: "Consciousness only dimly illuminates the . . . primitive elements in our experience" (*PR*, 162). The "primitive elements" in experience are those that enter through our (nonsensory) prehension. Because they occur before consciousness arises, they are generally only illuminated dimly, if at all.

The reason why we are not generally conscious of the "physical world" (or "nature") as filled with intrinsic value is that it is only in nonsensory prehension that we experience other things *as value-laden.* For example, we directly prehend our prior occasions of experience, and "remember" them with their joys, sufferings, and desires. We also directly experience, if in more blurred fashion, our bodily members, feeling their excitement, their enjoyment, their suffering, their thirst and hunger. If we would generalize this knowledge of what the "physical world" is like, we would not think of it as devoid of values, but as a throbbing multiplicity of energetic, passionate, appetitive events striving for, and realizing, values.

We have instead assumed, especially in the modern world, that it is our sensory perception, especially vision, that tells us what the physical world is really like. This assumption has had such fateful consequences because, in sensory perception, the element of value is virtually lost: what was received in prehension is transmuted, and the transmutation involves playing down the subjective, emotional nature of the data and playing up the objective, purely geometrical aspects.

In Whitehead's alternative account, "the primitive element [in experience] is *sympathy,* that is, feeling the feeling *in* another and feeling conformally *with* another" (*PR*, 162). Besides thereby reminding us that we directly know, through prehensions of our own bodies, that the world is comprised of beings with which we can sympathize, and that sympathy is our natural response,

Whitehead's account is designed to remove the tendency produced by consciousness to see ourselves as cut off from the world. Whitehead sees the task of philosophy to be "the self-correction by consciousness" of the tendency of consciousness to obscure "the external totality from which [the individual] originates and which it embodies." This self-correction involves the reminder that the individual, deep down, "has truck with the totality of things" (PR, 15). A central purpose of Whitehead's cosmology is to remind us of what we essentially are, below the superficialities of consciousness and sensory perception, with the hope that "the intellectual insight" will be converted "into an emotional force correct[ing] the sensitive experience in the direction of morality" (PR, 15). An epistemological revolution can help promote a moral revolution.

An Ecological God

A complete account of Whitehead's ecological worldview would require an account of his view of the God-world relation. Because there is no room for such an account here, however, I will simply mention the main points that would be included in such an account.

At the center of the difference between Whitehead's God and that of traditional theism is his dictum that "God is not to be treated as an exception to all metaphysical principles," but as "their chief exemplification" (PR, 343). On the one hand, God enters into all other actualities: "The world lives by its incarnation of God in itself" (RM, 149). On the other hand, all other actualities enter into God: "It is as true to say that the World is immanent in God, as that God is immanent in the World" (PR, 348). Because of this mutual influence, Whitehead's doctrine of God is sometimes called "panentheism." Unlike pantheism, God is not simply identical with the world; unlike traditional theism, God could not live apart from a world. God is essentially the soul of the world.[14] God's power is not coercive but persuasive, which means that God cannot unilaterally determine what happens in the world. The idea of God's goodness and love for the world is, accordingly, not undermined by the problem of evil.[15] This idea of divine power, furthermore, does not support the complacent belief that, if things get bad enough, God can intervene to save us.

Besides not being undermined by belief in this God, ecological commitment is actually reinforced. Our sense of the importance of our actions in relation to other creatures is reinforced by the idea that God is immanent in all beings, so that each species is a unique mode of divine presence (as Thomas Berry says), and by the idea that all creatures contribute value to the divine life. Finally, against an ethical skepticism that suspects that terms such as "good" and "right" are purely emotive, with no real meaning, the panentheist can reply that that which is right in any situation is that which God, as the all-inclusive sympathetic participant, prefers.[16]

Biological Egalitarianism

I come now to the issue mentioned at the outset: whether Whitehead's cosmology, besides being deeply ecological in the above-mentioned ways, can also, in some sense, affirm the idea that has been the most controversial feature of the position to which the label "deep ecology" is usually applied, the idea of biological egalitarianism. I suggest that it can, and in so doing can remove the major reason why this idea has been so divisive. The basis for the reconciliation is the recognition that, when Whiteheadians affirm and Naessians deny that various species of life have different degrees of "intrinsic value, the term "intrinsic value" is being used differently.

Naess has admitted that his intuition about biological egalitarianism is problematic. After describing his philosophy as "ecosophy," meaning "insight directly relevant to action," he has said that one cannot really act in harmony with statements about biological equality.[17] He has admitted, furthermore, that he does not know "how to work this [intuition] out in a fairly precise way" (ECL, 173). Two of his most careful attempts, evidently, are that "there is a value inherent in living beings which is the same value for all" (ECL, 168) and that "every living being is equal to all others to the extent that it has intrinsic value" (ECL, 184). Note that Naess seems to use "intrinsic" and "inherent" value interchangeably. What he and his followers mean by these terms is different from the meaning given to "intrinsic value" in my Whiteheadian discussion above. To minimize confusion, I suggest that the term "intrinsic value" henceforth be used for the Whiteheadian meaning, the term "inherent value" for the Naessian meaning.

The difference between the two meanings involves what the

contrast is *with*. The *intrinsic* value of something (in Whiteheadian discussions) stands in contrast with its (extrinsic) value for others—*any* others, be they plants, animals, humans, or God. The "other" can even be a later member of the same enduring individual: for example, a prior moment of my own experience, which at the time had intrinsic value for itself, now has extrinsic value for my present experience. The *inherent* value of something (in Naessian discussions) *stands in contrast solely with its (perceived) value for human beings*. That this is so is shown by the first point in an eight-point platform of the deep ecology movement written by Naess and George Sessions:

> The well-being and flourishing of human and nonhuman life on Earth have value in themselves (synonyms: intrinsic value, inherent value). These values are independent of the usefulness of the nonhuman world for human purposes.[18]

Accordingly, *"inherent value" in the Naessian sense includes most of what counts as "extrinsic value" in the Whiteheadian usage.* In particular, an organism's inherent value includes its *ecological* value, its value for sustaining the ecosystem. Because the only kind of value that is excluded from inherent value is the *(perceived) value of something for human beings,* then some other kinds of extrinsic value would also be included in inherent value, such as the aesthetic and instrumental values that things may have for nonhuman animals. For the sake of this discussion, however, we can focus on ecological value as the most important type of extrinsic value (in the Whiteheadian sense) that is included in inherent value (in the Naessian sense).

Another contrast between Whiteheadian "intrinsic value" and Naessian "inherent value" is that the former applies only to individuals, while the latter, because it includes both intrinsic and extrinsic value (in the Whiteheadian sense), applies to *species* as well as individuals.

The central implication of this terminological discussion is that *a rough equality in the inherent value of the various species results from an inverse relation that exists, in general, between intrinsic value and ecological value.* That is, those species whose (individual) members have the *least intrinsic value,* such as bacteria, worms, trees, and the plankton, have the *greatest ecological value:* without them, the whole ecosystem would collapse. By contrast, those species whose members have the *greatest intrinsic value* (meaning the richest experience and thereby the most value for themselves), such as whales, dolphins, and primates, have the *least ecological value*. In the case of human primates, in fact, the ecological value is negative: most of the other forms of life would be better off, and the ecosystem as a whole would not be threatened, if we did not exist. In any case, assuming that this inverse correlation generally obtains throughout the ecological pyramid, we can say that all forms of life have, roughly, the same inherent value, which is the distinctive point of egalitarian deep ecology.

Recognition that the *total* inherent value of anything includes both its intrinsic and its ecological values provides a basis for reconciling the erstwhile conflict between deep ecologists and advocates of the "land ethic," on the one hand, and humanitarians and animal liberationists, on the other.

Deep ecologists and land ethicists have been focusing primarily on ecological value. Given that focus, they rightly see that those species at the base of the ecological pyramid—such as the worms, the trees, the bacteria, the plankton—are vital. If these thinkers focus *exclusively* on ecological value, they may see concern for the liberation of humans and other mammals from suffering as diversionary or worse. The concern with individuals, as opposed to species, will seem misguided. The concern to protect humans from premature and massive death may even seem counterproductive, especially if the exclusive focus on ecological value leads to misanthropy. The charge of "speciesism" may be leveled.

Animal liberationists and humanitarians, on the other hand, focus primarily upon intrinsic value and therefore primarily upon individuals. Given this focus, animal liberationists rightly see that, among the nonhuman forms of life, the higher animals, especially mammals, have the greatest capacity for intrinsic value, and thereby the greatest capacity to suffer and to have their potentials for self-realization thwarted. One's ethical activity with regard to the nonhuman world is, accordingly, most appropriately directed toward preventing the suffering and ensuring the flourishing of the higher animals. Humanitarians see that the rights of our fellow human beings to live without unnecessary suffering and without unnecessary restraints on their flourishing should especially be protected. If these animal and human liberationists focus *exclusively* on intrinsic value, they may regard the efforts of deep ecologists as diversionary or worse. Seeing the misanthropy of some of them, they may mutter "ecological fascism."

Once we see, however, that the total *inherent* value of things, the total value things have *in themselves*, includes both their intrinsic value and their ecological value, and that these values generally exist in inverse proportion to each other, we can see these two ethical concerns as complementary, not conflictual. Both concerns are valid and need to be addressed simultaneously. Those who wish to focus their energies on one can do so wholeheartedly, knowing that other thinkers and activists are focusing their energies on the other.

The panentheistic perspective underwrites this "both/and" attitude. On the one hand, God (by hypothesis) has spent billions of years coaxing along a universe that could bring forth life and presumably does not wish the whole enterprise, on this planet at least, to come to an end, or at least to be drastically reduced, billions of years prematurely. If the divine perspective finally defines what is right, it is right to try to preserve the ozone shield, to prevent severe global warming, to preserve biodiversity, and to do whatever else is necessary to protect the integrity of the global ecosystem. On the other hand, God has not only brought forth life, but has continually evoked *increasingly complex* species of life, whose individual members are capable of increasingly richer intrinsic value and are thereby capable of contributing more value to the universe as a Whole, to God. Because God not only enjoys the enjoyments of the various creatures, but also suffers their sufferings, God presumably wants the suffering of the creatures minimized, their flourishing maximized. It is right, then, to give special attention to preventing suffering and enabling flourishing, especially with regard to those creatures who can suffer the most, and whose potentials can be most severely thwarted. The Whiteheadian God, accordingly, is a deep ecologist, but one whose deep ecology includes animal and human liberation.

Notes

1. Alfred North Whitehead, *Religion in the Making* (1926; reprint, Cleveland, Ohio: World Publishing Co., 1960), 50; hereafter, *RM*, with page references cited in the text.

2. For a thorough history of the term and controversies surrounding it, see Warwick Fox, *Toward a Transpersonal Ecology: Developing New Foundations for Environmentalism* (Boston: Shambhala, 1990).

3. Ibid., 75.

4. As Fox points out, George Sessions, in working out a classification scheme, listed the animal liberation movement as a form of shallow, rather than deep, ecology; see ibid., 65, 66.

5. Fox, after having long been an advocate of deep ecology, has decided that the term should be dropped. In his helpful discussion, he distinguishes three meanings of "deep ecology": the formal, the philosophical, and the popular. The *formal* meaning refers to

asking ever deeper questions until one gets down to fundamental assumptions (92, 125–27). Fox rightly argues that this meaning is untenable (131–41). The *popular* sense involves a biocentric approach, according to which all forms of life are respected as having value in themselves, and in which some kind of biospherical egalitarianism is accepted (114–17). While pointing out that this popular meaning is "the one by which the term *deep ecology* is by far the most widely known," Fox believes that it contains nothing distinctive in relation to most other ecophilosophies (118). This leaves the *philosophical* meaning, which refers to "Self-realization," and which involves "the realization of a more and more expansive sense of self," so that one identifies with more and more of the world (106). Fox believes that this "constitutes the essence of what is tenable and distinctive about the deep ecology approach to ecophilosophy" (118). On this basis, Fox suggests replacing the term with "transpersonal ecology." My decision to use "deep ecology" positively reflects a different assessment of the "popular" meaning of the term. The term "deep" popularly connotes the idea that inherent value goes all the way down, that there is no line below which we can rightly treat things as simply means to human ends. This point (which I have called deep ecology$_b$) does distinguish this approach from that of many ecophilosophies. And the insistence on biospherical egalitarianism (deep ecology$_e$), especially as interpreted by Devall and Sessions, has even more clearly set deep ecology apart. Because I consider these two points both distinctive and (as reinterpreted in the present essay) true and valuable, I am happy to associate the Whiteheadian approach with the term.

6. From the Whiteheadian side, see John B. Cobb, Jr., "Ecology, Science, and Religion: Toward a Postmodern Worldview," *The Reenchantment of Science: Postmodern Proposals*, ed. David Ray Griffin (Albany: State University of New York Press, 1988), 99–114, and Herman E. Daly and John B. Cobb, Jr., *For the Common Good: Redirecting the Economy Toward Community, the Environment, and a Sustainable Future* (Boston: Beacon Press, 1989), 377–78, where agreement with deep ecologists is expressed on all points except that of equality of intrinsic value. This contrast of Whiteheadian philosophy and deep ecology reflects the previous exclusion of Whiteheadians from the ranks of deep ecologists, on the grounds of "fail[ing] to meet the deep ecology norm of 'ecological egalitarianism in principle,'" by Bill Devall and George Sessions in *Deep Ecology: Living as if Nature Mattered* (Salt Lake City, Utah: Peregrine Smith, 1985), 236. It is interesting to note that Arne Naess himself, as shown even by a statement quoted by Devall and Sessions (ibid., 225), had *included* Whitehead among the deeply ecological philosophers.

7. Alfred North Whitehead, *Process and Reality: An Essay in Cosmology* (1929), corrected edition, ed. David Ray Griffin and Donald W. Sherburne (New York: Free Press, 1978), 167; see also xiii, 29, and 309; hereafter, *PR*, with page references cited in the text.

8. Alfred North Whitehead, *The Function of Reason* (1929; reprint, Boston: Beacon Press, 1958), 30–31.

9. Alfred North Whitehead, *Science and the Modern World* (1925; reprint, New York: Free Press, 1967), 93; hereafter, *SMW*, with page references cited in the text.

10. Alfred North Whitehead, *Modes of Thought* (1938; reprint, New York: Free Press, 1966), 110.

11. Charles Hartshorne has used the term "panpsychism," and, more recently, "psychicalism." See "Physics and Psychics: The Place of Mind in Nature," in *Mind in Nature: Essays on the Interface of Science and Philosophy*, ed. John B. Cobb, Jr. and David Ray Griffin (Washington, D.C.: University Press of America, 1977), 89–96, at 91. Those terms, however, tend to suggest the higher form of experience that only animal psyches can enjoy. They also tend to suggest that the basic units of the world are enduring things rather than momentary experiences. Hartshorne has recently said that he sees advantages in the term "panexperientialism." See "General Remarks," in *Hartshorne, Process Philosophy and*

Theology, ed. Robert Kane and Stephen H. Phillips (Albany: State University of New York Press, 1989), 181.

12. See Charles Hartshorne, "The Compound Individual," in *Philosophical Essays for Alfred North Whitehead*, ed. Otis H. Lee (New York: Longmans, Green, 1936), 193–220.

13. Alfred North Whitehead, *Adventures of Ideas* (1933; reprint, New York: Free Press, 1967), 204; hereafter, *AI*, with page references cited in the text.

14. On panentheism, see Charles Hartshorne and William Reese, eds., *Philosophers Speak of God* (Chicago: University of Chicago Press, 1953), vii, 1–57, 233–334, 499–514. On God as "World Soul," see Hartshorne, *Omnipotence and Other Theological Mistakes* (Albany: State University of New York Press, 1984), 59, 134–35. Whitehead said that God "is not the world, but the valuation of the world" (*RM*, 95, 152).

15. See David Ray Griffin, *God, Power and Evil: A Process Theodicy* (Philadelphia: Westminster Press, 1976; Lanham, Md.: University Press of America, 1991), and *Evil Revisited* (Albany: State University of New York Press, 1991).

16. David Ray Griffin, "The Holy, Necessary Goodness, and Morality," *Journal of Religious Ethics* 8, no. 2 (Fall 1980): 330–49.

17. Arne Naess, *Ecology, Community, and Lifestyle: Outline of an Ecosophy* (Cambridge: Cambridge University Press, 1989), 33, 174; hereafter, *ECL*, with page references cited in the text.

18. This platform was first published in George Sessions, *Ecophilosophy VI* (1984) and has been republished in Devall and Sessions, *Deep Ecology*, 70, and in Fox, *Toward a Transpersonal Ecology*, 114–15.

Deep Ecology as Worldview

George Sessions
Sierra College

Introduction

ARNE Naess maintains that deep ecology is a total view—a worldview—which involves not only our way of acting in daily life but also our most fundamental intuitions about our relationship to nonhuman nature, our basic values, and our beliefs and feelings about what is of ultimate importance in life. In this respect, he has claimed, deep ecology has a religious component.[1] Since the essays in the present volume are weighted toward the discussion of world religions and ecological views, I would like to pay special attention to the relation of religious beliefs to the ecological crisis and to the religious aspects of deep ecology.

Certainly, the early founders of ecocentrism and environmentalism in America, and the forerunners of the modern deep ecology movement, Henry David Thoreau and John Muir, were both deeply religious men whose ecological views were throughly enmeshed with their pantheistic identification of God with nature—with nature as sacred.[2] And UCLA historian Lynn White, Jr., whose 1966 paper touched off the wide-ranging controversy over Christian anthropocentrism and the ecological crisis, argued that "since the roots [of the ecological crisis] are so largely religious, the remedy must also be essentially religious, whether we call it that or not. We must rethink and refeel our nature and destiny." White claimed that:

> Especially in its Western form, Christianity is the most anthropocentric religion that the world has seen. . . . Christianity, in absolute contrast to ancient paganism and Asia's religions . . . not only established a dualism of man and nature but also insisted that it is God's will that man exploit nature for his proper ends.

White's religious solution was to return to the views of Saint Fran-

cis. Francis "tried to substitute the idea of the equality of all creatures, including man, for the idea of man's limitless rule of
creation." I will also be concerned to examine White's provocative
claim that "no new set of basic values has been accepted in our
society to displace those of Christianity."[3] In reply to a question
after a lecture at UC Davis in the early seventies, White remarked
somewhat jokingly that "with one stroke [the 'Historical Roots'
paper], I founded the theology of ecology."

I would also like to call attention to Max Oelschlaeger's recent
claim that, of all the institutions of modern liberal-industrial society, it is perhaps only religion and the church which have the
power to move modern societies beyond the dominant modern
worldview—the worldview of economic man and of life as pro-
fane—to an old/new sense of the sacredness of all life on earth.[4]

In addition, I would like to point to the analysis of contemporary worldviews or ideologies developed by two political scientists,
Donald Michael and Walter Truit Anderson. They claim that the
ideologies by which most contemporary people live are: the
American economic growth "story"; Christian and Islamic funda-
mentalism; Marxism; and further, that all these ideologies fail to
address the significant issues of the postmodern world.[5] Now that
Marxism is collapsing, many countries are rushing to embrace
the American capitalistic economic growth story, a story which,
along with industrial Marxism, is incompatible with ecologically
long-range sustainable futures.[6] What the "new world order"
(proclaimed by some to be emerging) amounts to is the intertwined octopus of an international corporate power elite, together
with interlocking international economic markets and bureaucracy, which has no allegiance to any country or nation, as the
working classes of America and the world are now beginning to
realize to their dismay. What we are ensnared in is not only the
destruction of the earth, but a further continuation of what the
poet Archibald McLeish called "the diminishment of man." I am
reminded of Loren Eiseley's remark, in discussing Robinson Jeffers's poetry: "Men feel, in growing numbers, the drawing of a
net of dependency against which something wild in their natures
still struggles as desperately as trapped fish in a seine."[7]

Michael and Anderson also point out that two Western postmodern worldviews are emerging: the "green story" and the "new
paradigm story." Given their description of these two "stories," I
would tend to identify them (with some important reservations
and qualifications), respectively, with the deep ecology movement
and the New Age movement. Some confusion arises here since the

deep ecology movement is also referred to as a "new paradigm" position.[8] Many people, including the media, fail to distinguish between the deep ecology and the New Age movements, but the differences are crucial, as I shall attempt to show, and are of special importance as the religions and churches of the world attempt to revise their theologies and teachings to reflect an ecological worldview and enter the arena of environmental activism.

The Deep Ecology Movement

Viewed historically, contemporary ecological consciousness began to arise in the United States after World War Two partly as a result of the publication of the ecologist Aldo Leopold's ecocentric "land ethic" (*A Sand County Almanac,* 1949). As a philosophical-social-political movement, ecocentric environmentalism can be dated from the publication of Rachel Carson's *Silent Spring* in 1962. Carson's total worldview and intimate combination of theory and practice marked the beginning of the so-called Age of Ecology and a period of intense philosophical and religious questioning of the fundamental assumptions, belief systems, values, and goals of Western culture. While a number of warnings of impending environmental disaster were made during the forties and fifties, Carson's efforts marked the beginning (together with David Brower's revolutionary leadership of the Sierra Club in the late fifties and sixties) of radical environmental activism (eliciting, as it did, major counterattacks from the chemical industry and the U.S. Department of Agriculture), thus triggering the "long range, international Deep Ecology movement."[9] As Naess points out: "classical nature conservation did not include fighting the power-centers which were pushing mindless 'development.' The environmental fight, from 1963–68, in California (and the U.S. generally) inspired the rest of the world. The 1972 United Nations Conference in Stockholm was the first acknowledgement by the establishment of social and political environmental conflicts." After the mainstream environmental organizations again tended to become timid, bureaucratic, and cozy with the power establishments in the seventies and eighties, the radical environmental fight was taken up by such groups as Greenpeace and Earth First![10]

In many ways, the Age of Ecology confirmed the earlier ecological insights and social criticism of Thoreau, Muir, D. H. Lawrence,

Robinson Jeffers, Aldous Huxley, and Leopold. Further roots of modern ecological consciousness and the deep ecology movement can be traced back to the ecocentric religions and ways of life of primal peoples around the world; Saint Francis; the Romantic countercultural movement with its roots in Spinoza; Taoism; and the Zen Buddhism of Alan Watts and Gary Snyder (which influenced many professional ecologists). Two best-selling books which brilliantly summarize much of the questioning and social-ecological criticism of the sixties are Charles Reich's *The Greening of America* (1970) and Theodore Roszak's *Where the Wasteland Ends* (1972).[11]

A distinguished Norwegian academic philosopher, Naess coined the term "deep ecology" and has been the leading philosophical articulator of the position. The Californian Pulitzer Prize–winning poet and essayist, Gary Snyder, also worked out a unique deep ecological worldview during the sixties which emphasized an ecocentric bioregional-reinhabitory approach. Naess and Snyder are the two most influential international exponents of the deep ecology movement.[12]

The way Naess articulates (or describes) his deep ecology position has not remained unchanged since the early seventies; at least three different and separate characterizations have been given: (1) Naess's early *description* of the beliefs, attitudes, and life-styles of supporters of the deep ecology movement in his 1972 Bucharest paper; (2) the development, during this period of the late sixties and early seventies, of Naess's own personal philosophical-ecological worldview (which he calls "Ecosophy T") based upon the ultimate norm "Self-Realization!" and involving the "wide-identification" thesis and a blending of the systems of Spinoza and Gandhi; and (3) the development, in 1984, of a deep ecology platform together with a supporting "apron diagram" with four levels. These levels illustrate the relation of ultimate religious and philosophical commitments and belief systems (Level 1) to the platform of deep ecology (Level 2), and further, how the first two levels relate logically to concrete ecological decisions and actions (Levels 3 and 4). As a worldwide social movement, the international deep ecology movement is best characterized by the deep questioning process, the deep ecology platform and the apron diagram, and the life-styles and ecological social-political actions which tend to follow from the platform.[13]

The 1972 Bucharest paper was largely a sociological descriptive account of what Naess saw as an international philosophical-social movement which had arisen during the sixties, based upon the experiences of field ecologists such as Carson and others who were closely associated with wild nature. These experiences produced both scientific conclusions and deep ecological intuitions which were amazingly similar all over the world. These included an awareness of the internal interrelatedness of ecosystems and the individuals comprising them; ecological egalitarianism (the same "right" of all species to live and flourish); an appreciation of ecological diversity, symbiosis, and complexity (and a humble awareness of our profound "human ignorance of biospherical relationships"); an anti social-class posture; together with the principles of local autonomy and decentralization. The ecological field-worker, Naess pointed out, "acquires a deep-seated respect, or even veneration, for ways and forms of life." As part of this *description,* Naess said that:

> the significant tenets of the Deep Ecology movement are clearly and forcefully *normative* ... insofar as ecology movements deserve our attention they are ecophilosophical rather than ecological. Ecology is a limited science which makes use of scientific methods. Philosophy is the most general forum of debate on fundamentals, and political philosophy is one of its sub-sections.[14]

Ecosophy T Naess calls the ecosophy he "feels at home with," which he has been developing and refining since the late sixties. There is an emphasis on the need for thinking in terms of norms. It starts with one ultimate (fundamental) philosophical norm "Self-Realization!" from which he derives various subnorms such as "Self-realization for all beings!," "No exploitation!", "No class society!", "Maximum complexity!", "Maximum diversity!", and "Maximum symbiosis!" That is, he derives these *norms* from "Self-Realization!" plus an important set of descriptive sentences. He insists that an articulated ecosophy needs a host of descriptions (or "hypotheses") concerning how the world actually is. What Naess means, in part, by "Self-Realization" is the universe (nature, the Tao) and all the individuals (human and nonhuman) of which it is comprised, realizing itself. Following the insights of Gandhi and Spinoza, human individuals attain personal self-realization and psychological-emotional maturity when they progress from an identification with narrow ego, through identification with other humans, to a more all-encompassing identification of their "self" with nonhuman individuals, species, ecosystems, and with the ecosphere itself. This process of "wide identification" Naess takes to be a process of the *development* of the "ecological self."[15]

The *1984 Deep Ecology Platform*, which characterizes deep ecology as a contemporary international ecophilosophical-social-political movement, came about, I think, as a result of the realization by Naess that some of the characteristics listed in the 1972 Bucharest paper were too specific in that they did not allow for full cultural diversity (such as the principles of local autonomy and decentralization), and further, that they included fundamental ecophilosophic beliefs (e.g., the doctrine of "internal relations" in characterizing ecosystems) not necessarily held by all suporters of the deep ecology movement. Ecosophies, while comprising all four levels of the "apron diagram," nevertheless had to be conceptually separated from more philosophically neutral general beliefs and attitudes shared by most, or all, supporters of the deep ecology movement (the Level 2 platform), thus allowing for, and promoting, a diversity of widely differing ecosophies (such as Spinozist, Christian, Muslim, Buddhist, and so on). An "anti social-class posture," while held by most supporters, is not specifically an *ecological* issue, as will be discussed below, and so is not included in the platform. The platform is essentially a statement of philosophical and normative ecocentrism together with a call for environmental activism.

The statement of the Deep Ecology Platform is as follows:

1. The well-being and flourishing of human and nonhuman life on earth have value in themselves (synonyms: intrinsic value, inherent value). These values are independent of the usefulness of the nonhuman world for human purposes.

2. Richness and diversity of life forms contribute to the realization of these values and are also values in themselves.

3. Humans have no right to reduce this richness and diversity except to satisfy *vital* needs.

4. The flourishing of human life and cultures is compatible with a substantial decrease of the human population. The flourishing of nonhuman life requires such a decrease.

5. Present human interference with the nonhuman world is excessive, and the situation is rapidly worsening.

6. Policies must therefore be changed. These policies affect basic economic, technological, and ideological structures. The resulting state of affairs will be deeply different from the present.

7. The ideological change is mainly that of appreciating *life quality* (dwelling in situations of inherent value) rather than adhering to an increasingly higher (materialistic) standard of living. There will be a profound awareness of the difference between big and great.

8. Those who subscribe to the foregoing points have an obligation directly or indirectly to try to implement the necessary changes.[16]

In his paper "Deep Ecology and Lifestyle" Naess also attempted to describe the life-styles of supporters of the deep ecology movement in terms of *tendencies* toward (1) using simple means; (2) anticonsumerism; (3) efforts to satisfy vital needs rather than desires; (4) going for depth and richness of experience rather than intensity; (5) attempts to live in nature and to promote community rather than society; (6) appreciation of ethnic and cultural differences; (7) a concern about the situation of the Third and Fourth Worlds and an attempt to avoid a standard of living too much different from and higher than the needy (global solidarity of life-style); (8) appreciation of life-styles which are universalizable, which are not blatantly impossible to sustain without injustice toward fellow humans or other species; (9) appreciating all life forms; (10) a tendency toward vegetarianism; (11) protecting wild species in conflicts with domestic animals; (12) efforts to protect local ecosystems; and (13) acting nonviolently.[17]

Deep ecology has sometimes been criticized as a social movement for not being sufficiently concerned with issues of social justice. This is largely a matter of emphasis and priorities and of sorting out *ecological* from other issues. Naess recently pointed out that the international Green movement is comprised of three movements: (1) the peace movement; (2) the social justice movement; and (3) the ecological movement, together with the goal of "wide" ecological sustainability. Naess claims that:

> Considering the accelerating rate of irreversible ecological destruction worldwide, I find it acceptable to continue fighting ecological unsustainability, whatever the state of affairs may be concerning the other two goals of green societies. I find this to be so, even in spite of the completely obvious requirement that there needs to be significant progress towards the goals of the peace and the social justice movements in order fully to reach ecological sustainability.[18]

One of the major concerns of the deep ecology movement is to protect the biological integrity and evolutionary processes of the earth, and further, to promote and help implement the findings of the new field of conservation biology. This new discipline has discovered that greatly expanded wild areas with interconnecting corridors are needed to protect and restore biodiversity through-out the world.[19]

Some critics have also claimed that deep ecology is too con-cerned with philosophical fundamentals and individual life-styles, and, as a result, has failed to develop an effective political posi-tion.[20] Some people, by temperament or training (especially in our modernist economic-political society), want to ignore the deep questioning process (which uncovers and questions our most fun-damental philosophical assumptions, belief systems, and values) in their preoccupation with politics and their rush to bring about political change. What often happens, as a result, is that they re-main trapped in the assumptions of the very system they want to change, proposing social-political solutions which are often either trivial or counterproductive, or both (what Thomas Berry refers to as microphase solutions to macrophase problems). Similarly, without the deep questioning process, people remain mesmerized by the contemporary industrial vision of reality and fall prey to trendy superficial socially approved "ecological" changes in life-styles, such as *merely* recycling or buying "green" products, while disregarding, for example, the crucial need to drastically alter their high-consumption life-styles and their expectations concern-ing the number of children they plan to have (in general, their overall negative impact on the biosphere).[21] And those fixated at the political-legal level also often miss the spiritual and "paradigm shift" dimensions of the ecological revolution.

The Deep Ecology Platform is structured so that it leads logi-cally from philosophical-ecological norms to political activism. Naess stresses that points 6 and 8 of the platform imply ecological-political activism, and that social-political activism is *absolutely cru-cial*. The unique character of the deep ecology movement, he claims, is "activism on a 'spiritual' basis" (acting from the basis of a fundamental philosophic-religious ecosophy, and acting nonvio-lently). In order to avoid misunderstanding concerning the cri-tique of anthropocentric reform ("shallow") environmentalism, Naess also points out that the importance of the social-economic-political activism of mainstream environmental organizations and individuals is fully appreciated, whether or not they are supporters of the deep ecology movement: "the ecological frontier is long."[22]

An Historical Digression: Western Views of the Human-Nature Relationship

The well-known Australian philosopher and historian of ideas, John Passmore, published the first widely known book in ecophilosophy in 1974: *Man's Responsibility for Nature* (Naess's 1973 *Okologi, samfunn, og livsstil* was not available in English translation until 1989). Passmore's conclusions were essentially anthropocentric and unecological: for example, he defined "conservation" as "the saving of natural resources for later consumption."[23] The most enduring part of Passmore's book is the first two chapters which consist of a penetrating analysis of the development of the dominant Western human-nature views. He characterized these as (1) man as despot; (2) stewardship; and (3) man developing and perfecting nature positions.

Man as despot: In keeping with White's thesis, Passmore traced the view of man as reckless exploiter and subduer to the historically dominant interpretation of Genesis in the Old Testament.[24]

Man as steward: The idea of human "stewardship" of the earth Passmore traced to Plato and to the post-Platonic philosopher, Iamblichus. In trying to explain why man's immaterial and immortal soul would ever "immerse itself in matter," Iamblichus (third century A.D.) referred to a passage from Plato: "Man, they said, is sent to earth by God 'to administer earthly things,' to care for them in God's name."

A more recent version of the stewardship doctrine, Passmore claims, arose from seventeenth-century humanism and the statement by Sir Matthew Hale that "the end of man's creation was that he should be the viceroy of the great God of heaven and earth in this inferior world: his steward [sty-warden], *villicus* (farm manager), bailiff or farmer of this goodly farm of the lower world." As Passmore points out, "Man is still to think of himself, on Hale's view, as master over the world."[25]

Rene Dubos proposed a Christian stewardship position based upon the teachings of Saint Benedict, pointing out that the Benedictine order "actively intervened in nature" as farmers and builders. According to Dubos:

Saint Benedict believed that it was the duty of the monks to work as partners of God in improving his creation or at least in giving it a more human expression. Implicit in his writings is the thought that labor is like a prayer which helps in recreating paradise out of chaotic wilderness.[26]

The dominant image of both the humanist and Christian versions of stewardship is of the earth as a potential farm to be thoroughly humanized and altered from its wild state. As Paul Shephard points out:

> As agriculture replaced hunting and gathering it was accompanied by radical changes in the way men saw and responded to their natural surroundings . . . . [Agriculturalists] all shared the aim of completely humanizing the earth's surface, replacing wild with domestic, and creating landscapes from habitat.[27]

In the early seventies, White and Dubos debated the merits of Franciscan ecological equality versus Benedictine "man over Nature" stewardship.[28]

By the seventeenth century, anthropocentric Christian and humanist views were mutually reinforcing, and difficult to separate, which tends to support White's claim that "we continue to live today very largely in a context of Christian axioms." For example, White argued that our faith in perpetual progress derives from Judeo-Christian teleology which "helps to show what can be demonstrated on many other grounds: that Marxism, like Islam [and Enlightenment modernist ideologies such as Deweyian pragmatism and, indeed, the American economic growth story] is essentially a Judeo-Christian heresy."[29]

Man perfecting nature: This third position Passmore traced to the Stoic, Posidonius. As Passmore describes this view:

> man's responsibility is to perfect nature by cooperating with it [in the sense that] we speak, in this spirit, of an area still in something like its original condition as "not yet developed." To "develop" land, on this view, is to actualize its potentialities, to bring to light what it has in itself to become, and by this means to perfect it . . . . How is perfection to be judged: the presumption is still, in Aristotle's manner, that nature is at its best when it fulfills men's needs—that this, indeed, is its reason for existing, what its potentialities are for. So to perfect nature is to humanize it, to make it more useful for man's purposes, more intelligible to their reason, more beautiful to their eyes . . . . Man does not complete the universe simply by being in it . . . he helps to create it.[30]

The "man perfecting nature" position came to full flowering, according to Passmore, within the German idealist metaphysics of Fichte and Hegel, and was incorporated into the thinking of Karl Marx, Herbert Marcuse, Pierre Teilhard de Chardin, and Ian McHarg. Passmore claimed that, for Marcuse, there are:

> two kinds of mastery: a repressive and a liberating one. Man's relationship to nature, Marcuse is prepared to admit, must at first be repressive, but as he civilises nature, he at the same time liberates it, frees it, as Hegel also suggests, from its "negativity," its hostility to spirit . . . . So what is wrong with our treatment of nature is . . . that we have used it destructively, as distinct from seeking to humanize it, spiritualize it.

For Teilhard de Chardin, humans "must work *with* the world. They are the first beings sufficiently rational to see what nature, through gradual evolution, is doing, and sufficiently powerful to help it on its path towards that final consummation for which 'the whole creation groaneth and travaileth until now.'"[31]

Passmore endorsed a combination of the "stewardship" and the "man perfecting nature" models of the human-nature relationship as a sound contemporary environmental philosophy (which is similar to the "shallow" environmental approach, inspired by the founder of the U.S. Forest Service, Gifford Pinchot, and referred to as Resource Conservation and Development). The year after his book appeared, Passmore withdrew his endorsement of the anthropocentrism of both the "stewardship" and the "man perfecting nature" positions, claiming that:

> We do need a "new" metaphysics which is genuinely not anthropocentric . . . . The working out of such a metaphysics is, in my judgment, the most important task which lies ahead of philosophy.[32]

The New Age Movement

The New Age Movement presents itself as a postmodern spiritual worldview. But it often serves as a religious-philosophical ideology for those, especially in the genetic engineering, computer and space technology, and mass media industries, who see themselves as God's chosen vector for further industrial progress, human colonization of outer space, and the human takeover of the earth's evolutionary process. New Age is thus basically the antithesis of deep ecology.[33] The New Age movement manifests itself in two ways:

(1) as a "pop culture" movement associated with California and Shirley MacLaine; crystal and pyramid power; spiritual "channeling"; entreprenurial spiritual gurus with big fees, together with Eastern and Western versions of "pop psychology" and other manifestations of dubious spirituality. Several critics have taken "cheap shots" at deep ecology by attempting to associate it with the "pop culture" and other versions of the New Age movement: Alston Chase included both New Age (Marilyn Ferguson and Willis Harmon) and deep ecology proponents (White, Roszak, Snyder et al.) under the rubric of what he called the "California Cosmologists," claiming that these ideas were all spawned in California's "redwood think tanks"; the social ecologist Murray Bookchin (in an early tirade) claimed that deep ecology had "parachuted into our midst from the Sunbelt's bizarre mix of Hollywood and Disneyland."[34]

(2) As a more intellectual version deriving primarily from the writings of Buckminister Fuller and the Jesuit priest, Pierre Teilhard de Chardin. The technologist Fuller (in *An Operating Manual for Spaceship Earth*, 1971) proposed that humans are now in a position to take control of the biological systems of the earth. For Fuller, there is no human overpopulation problem. Engineers and computers, together with cybernetic and systems theory, can provide purely technological solutions to the world's problems.[35]

Teilhard de Chardin's theology, as we have seen, was heir to the "man perfecting nature" tradition, and is characterized by the fusion of Christian spirituality with biological evolution and a technological domination of the earth by humans. The Christian scholar Frederick Elder found Teilhard's theology to be "fiercely anthropocentric": Teilhard envisioned "man's evolutionary movement toward a point of complete humanization" of the earth. (Elder also criticized Christian theologians Herbert Richardson and Harvey Cox for looking forward with approval to the earth as a wholly artificial urban environment.) As a religious alternative to Christian anthropocentrism, Elder endorsed Loren Eiseley's ecocentric-ecological worldview.[36] Teilhardian scholar Conrad Bonifazi has described Teilhard's worldview:

> In response to the question, What is the earth? [Teilhard] would say, the earth is man! . . . In us, evolution may come to a halt, because we are evolution. . . . [Teilhard] envisages mankind, born on this planet and spread over its entire surface, coming gradually to form around

its earthly matrix one single, hyper-complex and conscious arch-molecule, co-extensive with the planet itself.[37]

Teilhard's "redemption theology" (as well as both the Judeo-Christian "man as despot" and "stewardship" positions) is derived from the story of the "fall" in Genesis when *both* humans and nature fell from divine grace; hence *both* humans and nature are in need of redemption. According to this story, humans partly redeem themselves in God's eyes by redeeming nature: converting wild ecosystems into a human artifact (farms, cities, tree farms, mining and managing the world as a human "resource" and, in other ways, making nature "useful" for humans).[38] At the beginning of the Enlightenment and the rise of the modernist worldview, Francis Bacon's vision of the "New Atlantis" and human technological mastery over nature was also justified on Judeo-Christian grounds of regaining command over a degraded nature lost at the time of the "fall." Bonifazi assures us that Teilhard was an optimistic thinker in this regard:

> [For Teilhard] apocalyptic despair of this world is overarched by hope of transformation of the whole of creation . . . our implicit destiny in the myth of the fall, with its ramifications in the natural world, is spelled out in the myth of the restored paradise . . . there is undiminished hope of nature's inclusion within the process of salvation.[39]

Other New Age writers have written in the "man perfecting nature" tradition. Peter Vajk claims that "should we find it desirable, we will be able to turn the Sahara Desert into farms and forests, or remake the landscape of New England . . . . We are the legitimate children of Gaia; we need not be ashamed that we are altering the landscapes and ecosystems of Earth."[40] The philosopher Henryk Skolimowski has also been inspired by Teilhard. For Skolimowski, the world is sacred (a sanctuary), but humans are the priests of the sanctuary: "The coming age is to be seen as the age of stewardship: we are here . . . to maintain and creatively transform, and to carry on the torch of evolution." Like Teilhard (and the New Age movement in general) Skolimowski sees humans as the culmination of the earth's evolutionary processes; correspondingly, he holds a graded hierarchy of intrinsic value with humans at the top of the pyramid:

> We cannot sustain all forms of life. Within the structure of evolution, the more highly developed the organism, the greater is its complexity

and its sensitivity and the more reason to treat it as more valuable and precious than others.[41]

Not only is this approach unecological, it is also not consistent with a scientific understanding of biological evolution. One is reminded of the note Darwin scribbled to himself in terms of thinking about evolution: "Never use the words *higher* and *lower*."[42]

While Anderson doesn't count himself as part of the New Age movement, he nevertheless writes in the "man perfecting nature" tradition when he says: "while most environmentalists are searching for ways to lessen human intervention in the natural world, I believe that intervention is, in a sense, human destiny, and that our task is to learn how we may sanely and reverently take responsibility for the global ecosystem and the course of evolution."[43] And in rejecting deep ecology and the deep questioning process, Anderson sees environmental issues as primarily political; he proposes that humans become the "business managers" of the earth's evolutionary processes.

Bookchin has criticized the New Age as well as the deep ecology movements. Bookchin's social ecology presents itself as a postmodern *ecological* worldview. But as Warwick Fox argues, social ecology (as the name suggests) all but ignores the major ecological issues, focusing instead mainly on issues of social justice.[44] This is explainable, in part, by the realization that Bookchin's position is an outgrowth of the Aristotelian-Hegelian-Marxian tradition and that it remains essentially within the parameters of the Enlightenment and the German idealist "man perfecting nature" visions (so characteristic of the New Age).

Robin Eckersley criticizes Bookchin for holding the view (like Teilhard) that humans, through their rationality, can determine the direction of evolution. Bookchin's position conceptually opens up wild nature (what he calls "first nature") to continued humanization and domestication by "second nature" (the domain of civilized urban culture). And, like Marcuse, the wild is "liberated" and made "free" when humans override natural spontaneous processes and "rationally" take control of the earth's evolutionary processes. According to Bookchin, humanity is an "agent for rendering evolution . . . fully self-conscious . . . [and] as rational as possible in meeting non-human and human needs." He recently concedes that some wilderness areas should be preserved free from major human intervention.[45]

The ecologist David Ehrenfeld has criticized Bookchin for his optimism, in the face of ecological realities, in proposing a toilless technological utopia for humanity. Ehrenfeld claims that "Bookchin and others like him have fled from reality to an altogether more soothing world of techno-pastoral dreams."[46]

Toward an Ecological Worldview and Ecologically Sustainable Societies

The leading environmental science textbook writer, G. Tyler Miller, saw things clearly in 1972 when he called for an end to the unecological Judeo-Christian/Greek humanist ideology of human domination, dominion, and control over the earth:

> Our task is not to learn how to pilot spaceship earth. [For Miller, the metaphor "spaceship earth" is itself an arrogant mechanistic misdescription of an organic earth.] It is not to learn how—as Teilhard de Chardin would have it—to "seize the tiller of the world." Our task is to give up our fantasies of omnipotence. In other words, we must stop trying to steer. The solution to our present dilemma does not lie in attempting to extend our technical and managerial skills into every sphere of existence. Thus, *from a human standpoint our environmental crisis is the result of our arrogance towards nature.*
>
> Somehow we must tune our senses again to the beat of existence, sensing in nature fundamental rhythms we can trust even though we may never fully understand them. We must learn anew that it is we who belong to the earth and not the earth to us. This rediscovery of our finitude is fundamental to any genuinely human future.[47]

A major breakthrough for Judeo-Christian views of dominion, and Teilhard's anti-ecological orientation came, in a dramatic and forthright move from the Catholic scholar (and president of the American Teilhard de Chardin Society) Thomas Berry. In 1982, Berry argued for a major ecological revision of Teilhard's theology, claiming that "the opinion is correct that Teilhard does not in any direct manner support the ecological mode of consciousness." Berry pointed out that:

> [For Teilhard] the sense of progress was irresistible . . . a world under rational control was the ideal to work toward . . . he fully accepted the technological and industrial exploitation of the planet as a desirable human activity . . . Teilhard is deeply involved in the total religious and humanistic traditions of the West out of which this exploitative attitude developed . . . Teilhard establishes the human as his exclusive

norm of values, a norm that requires the human to invade and to control rationally the spontaneities of nature . . . there is no question of accepting the natural world in its own spontaneous modes of being . . . this would be a treachery to the demands of the evolutionary process.

For Berry, Teilhard's position requires a totally different orientation to reflect an ecocentric-ecological worldview:

the evolutionary process [Berry claims] finds its highest expression in the earth community seen in its comprehensive dimensions, not simply in a human community reigning in triumphal dominion over the other components of the earth community. The same evolutionary process has produced all the living and non-living components of the planet.[48]

In claiming that Nature's spontaneous processes and modes of being must be respected, Berry's ecocentrism (and his more recent bioregionalism) merges with the deep ecological vision of Snyder when he wrote:

the unknown evolutionary destinies of other species are to be respected . . . what we envision is a planet on which [a much smaller] human population lives harmoniously and dynamically by employing a sophisticated and unobtrusive technology in a world environment which is "left natural."[49]

Contrary to the claims of some critics, the deep ecology-ecocentric position is not misanthropic but it does involve a major ecological reevaluation of "humanity's place in nature." For example, Berry holds that humans are a unique species as a result of our self-reflexive consciousness. We are the only species that can appreciate the main features of the cosmological and biological evolutionary processes. Naess also claims that we are "very special beings!" The problem, in Naess's view, is that we "underestimate our potentialities" both as individuals and as a species. Our ability to understand and identify with life on earth suggests a role primarily as "conscious joyful appreciator of this planet as an even greater whole of its immense richness." We can mature into being an "ecological self."[50]

As humanity moves toward the vision of ecologically sustainable societies, some thinkers are suggesting that evolutionary biology and cultural anthropology point to the significance of our physical-psychological-emotional genetic development as hunters and gatherers in wild natural environments. In terms of the development of Level 1 ecosophies, and following the lead of Naess's psychological analysis of the "ecological self," Warwick Fox's "transpersonal ecology," Paul Shepard (in *Nature and Madness*), and what Snyder calls a "depth ecology," Theodore Roszak has recently explored the groundwork for an "ecopsychology." Roszak claims, for instance, that "repression of the ecological unconscious is the deepest root of collusive madness in industrial society."[51] Similarly, Max Oelschlaeger has called for a postmodern return to "Paleolithic consciousness."[52] As is typical throughout the history of our species, major religious-philosophical paradigm shifts often result in new/old perspectives on what it is to be human.

Notes

1. See Stephan Bodian, "Simple in Means, Rich in Ends: Interview with Arne Naess" (1982), in *Environmental Philosophy: From Animal Rights to Radical Ecology*, ed. Michael Zimmerman et al. (Englewood Cliffs, N.J.: Prentice-Hall, 1993), 186.

2. For discussions of the religious-ecological views of Thoreau and Muir, see Max Oelschlaeger, *The Idea of Wilderness: From Prehistory to the Age of Ecology* (New Haven: Yale University Press, 1991), 133–204; Donald Worster, *Nature's Economy: A History of Ecological Ideas* (San Francisco: Sierra Club Books, 1977); and Michael P. Cohen, *The Pathless Way: John Muir and American Wilderness* (Madison: University of Wisconsin Press, 1984), chaps. 1, 6, and 7.

3. Lynn White, Jr., "The Historical Roots of Our Ecologic Crisis," in *Machina Ex Deo* (Cambridge: MIT Press, 1968), 93, 86, 93; for excellent discussions of the impact and implications of White's thesis, see Stephen Fox, *John Muir and His Legacy: The American Conservation Movement* (Boston: Little, Brown, 1981), 358–74; Roderick Nash, *The Rights of Nature: A History of Environmental Ethics* (Madison: University of Wisconsin Press, 1989), 87–120; and, especially, Roderick French, "Is Ecological Humanism a Contradiction in Terms?: The Philosophical Foundations of the Humanities Under Attack," in *Ecological Consciousness: Essays from the Earthday X Colloquium*, ed. J. Donald Hughes and Robert C. Schultz (Washington, D.C.: University Press of America, 1981), 43–66.

4. Max Oelschlaeger, "Caring for Creation: Religion in a Time of Ecological Crisis," in *After Earth Day: Continuing the Conservation Effort*, ed. Max Oelschlaeger (Denton: University of North Texas Press, 1992), 215–31.

5. Donald Michael and Walter Truit Anderson, "Now that 'Progress' No Longer Unites Us," *New Options* 33 (November 1987).

6. For a critique of industrial capitalism and socialism, see Andrew McLaughlin, *Regarding Nature: Industrialism and Deep Ecology* (Albany: State University of New York Press, 1993); see also Robin Eckersley, *Environmentalism and Political Theory: Toward an Ecocentric Approach* (Albany: State University of New York Press, 1992); Robert Paehlke, *Environmentalism and the Future of Progressive Politics* (New Haven: Yale University Press, 1989).

7. Loren Eiseley, Forward to David Brower, ed., *Not Man Apart: Lines from Robinson Jeffers* (San Francisco: Sierra Club Books, 1965).

8. See, for example, Fritjof Capra, "Deep Ecology: A New Paradigm," *Earth Island Journal* 2, no. 4 (Fall 1987): 27–30.

9. For discussions of these developments, see Stephen Fox, *John Muir and His Legacy*, 250–329; Michael P. Cohen, *The History of the Sierra Club: 1892–1970* (San Francisco: Sierra Club Books, 1988), esp. 187–394; Ralph Lutts, "Chemical Fallout: Rachel Carson's *Silent Spring*, Radioactive Fallout, and the Environmental Movement," *Environmental Review* 9 (1985): 210–25.

10. For an excellent critique of the reform environmentalism of the seventies and eighties, see Christopher Manes, *Green Rage: Radical Environmentalism and the Unmaking of Civilization* (Boston: Little, Brown, 1990), 45–65; see also Rik Scarce, *Ecowarriors: Understanding the Radical Environmental Movement* (Chicago: Noble Press, 1990); Dave Foreman, *Confessions of an Ecowarrior* (New York: Harmony Books, 1991).

11. For further discussions of the historical precedents leading up to the contemporary deep ecology movement, see Oelschlaeger, *The Idea of Wilderness;* Warwick Fox, *Toward a Transpersonal Ecology: Developing New Foundations for Environmentalism* (Boston: Shambhala Publications, 1990). Roderick Nash, *The Rights of Nature;* Roderick Nash, *American Environmentalism: Readings in Conservation History*, 3d ed. (New York: McGraw-Hill, 1990); Del Ivan Janik, "Environmental Consciousness in Modern Literature (Lawrence, Huxley, Jeffers, and Snyder)," in Hughes and Schultz, eds., *Ecological Consciousness*, 67–82; George Sessions, "Shallow and Deep Ecology: A Review of the Philosophical Literature," in Hughes and Schultz, eds., *Ecological Consciousness*, 391–462; George Sessions, "Ecological Consciousness and Paradigm Change" (1981), and Arne Naess, "Identification as a Source of Deep Ecological Attitudes," in *Deep Ecology*, ed. Michael Tobias (San Diego, Calif.: Avant Books, 1985), 28–44, and 256–70; George Sessions, "The Deep Ecology Movement: A Review," *Environmental Review* 11, no. 2 (1987): 105–25.

12. A succinct statement of Snyder's early deep ecological position occurs in Snyder, "Four Changes" (1969), modified and published in Gary Snyder, *Turtle Island* (New York: New Directions, 1974), 91–102; see also Gary Snyder, *The Practice of the Wild* (San Francisco: North Point Press, 1990); for recent commentary on Snyder, see Oelschlaeger, *The Idea of Wilderness*, 261–80; George Sessions, "Gary Snyder: Post-Modern Man," in *Gary Snyder: Dimensions of a Life*, ed. Jon Halper (San Francisco: Sierra Club Books, 1991), 365–70.

13. Warwick Fox, in *Toward a Transpersonal Ecology*, identifies deep ecology with the "wide-identification" thesis. This confuses Level 1 ecophilosophical commitments with the more characteristic Level 2 Deep Ecology Platform. Naess considers Fox's "transpersonal ecology" to be an important type of Level 1 ecosophy. The first two parts of Fox's book are an outstanding academic history and analysis of the development of the deep ecology movement, with exhaustive bibliographies.

Bill Devall and George Sessions, *Deep Ecology: Living as if Nature Mattered* (Salt Lake City, Utah: Peregrine Smith, 1985) is a semipopular exposition of deep ecology which unfortunately was hastily thrown together as a book from bits and pieces of previously published academic papers at the insistence of the publisher. It, too, misleadingly mixes Level 1 with Level 2 aspects of the apron diagram but contains a statement of the Deep Ecology Platform (pp. 69–73) and the apron diagram together with a discussion and diagram of Naess's Ecosophy T (appendix A, pp. 225–28).

Arne Naess, "The Deep Ecology Movement: Some Philosophical Aspects," (1983), reprinted in Zimmerman, ed., *Environmental Philosophy*, contains a statement of the Deep Ecology Platform and the apron diagram.

Arne Naess, *Ecology, Community and Lifestyle: Outline of an Ecosophy* (Cambridge: Cambridge University Press, 1989) is an authoritative academic treatment of the deep ecology position and contains a statement of the Deep Ecology Platform and a discussion of Ecosophy T. This book is a revised English translation of the 5th Norwegian edition of Naess, *Okologi, samfunn, og livsstil* (first published in 1973) which, in turn, arose out of a third expanded edition of a short work, *Ecology and Philosophy*, Naess had begun in the late sixties.

14. Lecture delivered to the Third World Future Research Conference in Bucharest in 1972 and published as "The Shallow and the Deep, Long-Range Ecology Movements: A Summary," *Inquiry* (Oslo) 16 (1973): 99. Naess claims that the original lecture notes were lost, but that they summarized part of an earlier version of *Okologi, samfunn, og livsstil.*

15. The term "ecological self" was coined and discussed in Arne Naess,"Self-Realization: An Ecological Approach to Being in the World," Roby Memorial Lecture (March 1986), Murdoch University, Western Australia: reprinted in *The Trumpeter: Canadian Journal of Ecosophy* 4, no. 3 (1987): 35–42.

16. There are extensive comments on each of the eight points of the platform, published along with the platform, in Devall and Sessions, *Deep Ecology;* Naess, "The Deep Ecology Movement;" Naess, *Ecology, Community, and Lifestyle* (see n. 13 above); see also Arne Naess, "The Encouraging Richness and Diversity of Ultimate Premises in Environmental Philosophy," *The Trumpeter* 9, no. 2 (1992): 53–60.

17. Arne Naess, "Deep Ecology and Lifestyle," in *The Paradox of Environmentalism,* ed. Neil Everndon, Symposium Proceedings, Faculty of Environmental Studies, May 1983 (Ontario: York University, 1984), 57–60. See also George Sessions, "Arne Naess and the Union of Theory and Practice," *The Trumpeter* 9, no. 2 (1992): 73–76 (special issue commemorating Arne Naess's eightieth birthday).

18. Arne Naess, "Politics and the Ecological Crisis," *Revision* 13, no. 3 (1991): 143; see also Arne Naess, "The Three Great Movements," *The Trumpeter* 9, no. 2 (1992): 85–86.

19. For recent discussions of conservation biology and the protection of biodiversity, see George Sessions, "Ecocentrism, Wilderness, and Global Ecosystem Protection," in *The Wilderness Condition: Essays in Environment and Civilization,* ed. Max Oelschlaeger (San Francisco: Sierra Club Books, 1992), 90–130; Curt Meine, "Conservation Biology and Sustainable Societies," in Oelschlaeger, ed., *After Earth Day,* 37–65; E. O. Wilson, ed., *Biodiversity* (Washington, D.C.: National Academy Press, 1988); R. Edward Grumbine, *Ghost Bears: Exploring the Biodiversity Crisis* (Washington, D.C.: Island Press, 1992).

20. See, for example, Andrew Dobson, "Deep Ecology," *Cogito* 3, no. 1 (1989): 41–46.

21. See Paul and Anne Ehrlich, *Healing the Planet: Strategies for Resolving the Environmental Crisis* (Reading, Mass.: Addison-Wesley, 1991), 1–14.

22. For Naess's recent views on ecological politics, see Naess, "Politics and the Ecological Crisis."

23. John Passmore, *Man's Responsibility for Nature: Ecological Problems and Western Traditions* (New York: Scribner's, 1974), 73.

24. Ibid., 3–27.

25. Ibid., 30.

26. Rene Dubos, *A God Within* (New York: Scribner's, 1972), 135–74; see also Rene Dubos, *The Wooing of the Earth* (New York: Scribner's, 1980).

27. Paul Shepard, *The Tender Carnivore and the Sacred Game* (New York: Scribner's, 1973), 237.

28. See Rene Dubos, "A Theology of the Earth," and Lynn White, Jr., "Continuing the Conversation," in *Western Man and Environmental Ethics,* ed. Ian G. Barbour (Reading, Mass.: Addison-Wesley, 1973); on 6 April 1980 a papal bull (issued by Pope Paul II) was announced in Assisi proclaiming Saint Francis to be the patron saint of ecologists, but, as White pointed out, Benedictine views were attributed to Francis: specifically that Francis "considered nature as a marvellous gift from God to humanity." White quipped, "What *will* become of the dogma of papal infallibility if this sort of bungling continues?!" (personal correspondence, 1 May 1980).

29. White, "Historical Roots of Our Ecologic Crisis," 85; see also George Sessions, "Ecocentrism and the Anthropocentric Detour," *Revision* 13, no. 3 (1991): 109–15.

30. Passmore, *Man's Responsibility for Nature,* 32–33.

31. Ibid., 35, 34.

32. John Passmore, "Attitudes toward Nature," in *Nature and Conduct*, ed. R. S. Peters (New York: Macmillan, 1975), 260; for an excellent critique of the Resource Conservation and Development position, see John Rodman, "Four Forms of Ecological Consciousness Reconsidered," in *Ethics and the Environment*, ed. T. Attig and D. Scherer (Englewood Cliffs, N.J.: Prentice-Hall, 1983), 82–92.

33. For a general introduction to the New Age movement, see Marilyn Ferguson, *The Aquarian Conspiracy: Personal and Social Transformation in the 1980's* (New York: St. Martin's Press, 1980); for an earlier version of this critique of New Age, see George Sessions, "Deep Ecology, New Age, and Gaian Consciousness," *Earth First! Journal* 7, no. 8 (1987): 27, 29–30.

34. Alston Chase, *Playing God in Yellowstone* (Boston: Atlantic Monthly Press, 1986), 344–62; for critiques of Chase's proposal for intensive wildlife management in Yellowstone National Park, see Dave Foreman, Doug Peacock, and George Sessions, "Who's 'Playing God in Yellowstone'?" *Earth First! Journal* 7, no. 4 (1986): 18–21; Holmes Rolston III, "Biology and Philosophy in Yellowstone," *Biology and Philosophy* 4 (1989); Murray Bookchin, "Social Ecology versus 'Deep Ecology': A Challenge for the Ecology Movement," *Green Perspectives: Newsletter of the Green Program* (Summer 1987), 3; Willis Harmon is reported now taking a more deep ecological approach to environmental issues.

35. For critiques of New Age cybernetic approaches to an ecological worldview, see Morris Berman, "The Cybernetic Dream of the Twenty-First Century," *Journal of Humanistic Psychology* 26, no. 2 (1986): 24–51; Bill Devall and George Sessions, "The Development of Natural Resources and the Integrity of Nature," *Environmental Ethics* 6, no. 4 (1984): 293–322.

36. Frederick Elder, *Crisis in Eden: A Religious Study of Man and the Environment* (Nashville, Tenn.: Abingdon Press, 1970), 16–17.

37. Conrad Bonifazi, "Teilhard de Chardin and the Future," paper read at Rice University, Houston, Texas, October 1968.

38. Judeo-Christian views of wilderness and primal peoples as evil and in need of "redemption" greatly influenced and rationalized the European invasion, settlement, and exploitation of the North American continent (and elsewhere throughout the world); this is documented in the opening chapters of Roderick Nash, *Wilderness and the American Mind*, 3d ed. (New Haven: Yale University Press, 1982).

39. Conrad Bonifazi, *The Soul of the World: An Account of the Inwardness of Things* (Washington, D.C.: University Press of America, 1978), 218–19, 220–21.

40. J. Peter Vajk, *Doomsday Has Been Cancelled* (Menlo Park, Calif.: Peace Publishers, 1978), 61.

41. Henryk Skolimowski, *Eco-Philosophy: Designing New Tactics for Living* (London: Boyers, 1981), 54–55, 83–84.

42. The Darwin quote appears in Nash, *The Rights of Nature*, 42.

43. Walter Truit Anderson, *To Govern Evolution: Further Adventures of the Political Animal* (New York: Harcourt Brace Jovanovich, 1987), 203; for ecological critiques of biotechnology and the human takeover of earth's evolutionary processes, see Jeremy Rifkin, *Algeny* (New York: Viking Press, 1983); Jeremy Rifkin, *Declaration of a Heretic* (Boston: Routledge & Kegan Paul, 1985).

44. Warwick Fox, "The Deep Ecology-Ecofeminism Debate and Its Parallels," *Environmental Ethics* 11, no. 1 (1989): 5–25; reprinted in Zimmerman, ed., *Environmental Philosophy*.

45. Robin Eckersley, "Divining Evolution: The Ecological Ethics of Murray Bookchin," *Environmental Ethics* 11, no. 1 (1989): 99–116; Murray Bookchin, *Remaking Society: Pathways to a Green Future* (Boston: South End Press, 1990), 204; for further criticism of Bookchin

and social ecology, see Kirkpatrick Sale, "Deep Ecology and Its Critics," *The Nation* 22 (14 May 1988): 670–75.

46. David Ehrenfeld, *The Arrogance of Humanism* (Oxford: Oxford University Press, 1978), 54, 127.

47. G. Tyler Miller, Jr., *Replenish the Earth: A Primer in Human Ecology* (Belmont, Calif.: Wadsworth, 1972), 53; see also G. Tyler Miller, Jr., *Living in the Environment: An Introduction to Environmental Science*, 6th ed. (Belmont, Calif.: Wadsworth, 1990).

48. Thomas Berry, *Teilhard in the Ecological Age* (Chambersburg, Pa.: Anima Books, 1982); quotes are from a selection from this book reprinted in Devall and Sessions, *Deep Ecology*, 143.

49. Snyder, "Four Changes."

50. See Thomas Berry, "The Viable Human," *Revision* 9, no. 2 (1987): 75–81, reprinted in Zimmerman, ed., *Environmental Philosophy;* Thomas Berry, *The Dream of the Earth* (San Francisco: Sierra Club Books, 1988); Arne Naess, "The Green Society and Deep Ecology" (The 1987 Schumacher Lecture), unpublished ms., 18; Arne Naess, "The Arrogance of Antihumanism?" *Ecophilosophy Newsletter* (1984), 8.

Michael Zimmerman favorably compares Naess's views with the views of Bookchin, but it is of course one thing to be primarily an "appreciator" of the biotic exuberance of the earth, quite another to promote major intervention and control of the evolutionary processes; see Zimmerman, "Deep Ecology, Ecoactivism, and Human Evolution," *Revision* 13, no. 3 (1991): 127.

In an otherwise powerfully written and provocative book, Vice President Gore seriously misrepresents the deep ecology position as being misanthropic by identifying it with personal statements made by Earth First! activists. Oddly enough, he claims that Naess "seems to define human beings as an alien presence on the earth" and as "having no intellect or free will with which to understand and change the [destructive anti-ecological] script we are following." One wonders how the Christian stewardship position he endorses (as partly a Level I ecosophy) differs, if at all, from the Level 2 Deep Ecology Platform; see Albert Gore, Jr., *Earth in the Balance: Ecology and the Human Spirit* (Boston: Houghton Mifflin, 1992), 216–18, 238–65.

51. Naess, "Self Realization: An Ecological Approach to Being in the World"; Warwick Fox, *Toward a Transpersonal Ecology;* Paul Shepard, *Nature and Madness* (San Francisco: Sierra Club Books, 1982); Snyder, *Practice of the Wild;* Theodore Roszak, *The Voice of the Earth* (New York: Simon & Schuster, 1992), 320.

52. Oelschlaeger, *The Idea of Wilderness*, 1–30, 320–53; see also Dolores LaChapelle, *Sacred Land Sacred Sex: Rapture of the Deep* (Silverton, Colo.: Flinn Hill Arts, 1988); Paul Shepard, "A Post-Historic Primitivism," in Oelschlaeger, ed., *The Wilderness Condition*, 40–89.

Ecological Geography

Thomas Berry
Riverdale Center for Religious Research

GEOGRAPHY is presented here as an integrating discipline that might contribute significantly toward a total earth science that seems to be needed in these times of such disturbed relations between the human and the biosystems of the planet. An integral functioning of the entire earth community is a condition for any sustainable mode of human presence upon the planet. This integral functioning must be effective in the existing order of things, not simply a theoretic vision. The human is a subsystem to the earth system. If the earth is not functioning properly, then there is no possibility of the human attaining any proper expression of itself.

While this relationship appears to be simple enough, it is surely among the most subtle adjustments that the planet has ever had to make. Bringing the human into being has been a daring venture from the beginning. The role of the human in relation to the other components of the earth community in its earlier phases was apparently quite clear. The intimacy, the sense of presence, the feeling for the numinous quality of the entire universe was all-pervasive. Only later, it seems, did the problematic aspects of human presence appear. Since the beginning of the Neolithic there have been superb theoretic presentations of what a valid context of human-earth relations might be, but these have seldom prevented the devastation of the various bioregions by their human inhabitants. The difficulty generally has been the effort of humans to achieve their own well-being at the expense of the surrounding natural regions.

The Chinese had a remarkable perception of the universe and the role of the human in the universe; yet in practice the Chinese were unable to carry on their cultivation of the land while preserving their forests. From their earliest period in North China, the Chinese started clearing the forests to increase the area for cultivation. Once this was begun, it was continued relentlessly throughout the centuries. Consequently, the hills were denuded of their vegetation and the soils eroded into the sea. This pattern was continued in other regions by human communities through the centuries as the wooded areas of the planet were cut down for farming or shelter or shipbuilding or firewood or some other human use. The land areas surrounding the Mediterranean were once fertile and forested lands. Yet even in classical times these regions were eroded and the soils began to wash into the sea. Europe lost much of its forest cover and its wildlife, especially in the postmedieval period when commerce increased and shipbuilding and mining and energy use were advanced. In Europe, however, farming methods took care of the fertility of the soil by fertilizing with animal waste, rotation of crops, and periodic leaving the land fallow. More recently immense problems have arisen as the natural systems of the earth were disrupted by a plundering industrial economy.

A certain ambivalence in human-earth relations came to manifest itself as humans assumed increasing control over the natural forces on the planet. There was a high esteem for the natural world as manifestation of the divine and as the primordial liturgy to be entered into through human rituals. Yet the natural world was seen also as a resource base for sustaining the physical and cultural well-being of the human. This ambivalence could be sustained by the earth in the earlier centuries. The planet was vast in relation to the number of humans and their limited ability to intrude on the functioning of the planet. Only in the past two centuries with the increase in human ability to plunder the planet has the stress between humans and the other components of the earth community become so acute as to be no longer tolerable.

To remedy this situation, a reexamination of the human presence upon the earth is causing serious foreboding throughout the human community. This reexamination is finding expression in movements sometimes referred to as environmental movements, sometimes as ecology movements. The former were focused more on the adjustment of the earth community to the needs of the human. The latter were concerned more with adjustment of the human to the needs of the earth community.

Prior to the emergence of the human, the earth, despite its volcanoes and hurricanes and tidal waves, its droughts and floods, its infestations and diebacks—despite all these destructive experiences, the earth did marvelously well in the larger arc of its development. That the number of species should be so vast and their interaction so intimate is the wonder of the earth. No other planet has, so far as we can determine, any indication of life, much less such magnificent forms or such diversity of life. Even after all these years of scientific inquiry into the structure of the planet and its life systems, our knowledge of the species inhabiting this planet is limited to something over a million and a half species, while we estimate that there are at least ten to twenty million species occupying the earth.

The interaction of these species has been enormously fruitful, although there have been both major and minor moments when extinctions have occurred. These occurred especially at the transition from the Paleozoic period to the Mesozoic period, some two hundred and twenty million years ago, and at the transition from the Mesozoic to the Cenozoic some sixty-five million years ago. At the present time another extinction is taking place that must be considered on this same scale. We are terminating the Cenozoic era. Our hope must be that we are entering into another creative period that might be identified as the Ecozoic era.

We are into a new biological situation, but, even more important, we are into a new planetary situation. Just how to describe this situation is difficult. While we generally use the terms "environmental" or "ecological," it might be more appropriate to deal with the situation in terms of a planet that has become dysfunctional because we do not have an integral sense of the earth or how it functions. We need a new study that might be designated, in the terms of Robert Muller, as a "Total Earth Science," a science that has so far never been properly identified as a special field of study.

While we have the term "cosmology" that deals with the entire physical universe, the term "geology" that deals with the physical nonliving structure of the earth, and the term "biology" that deals with the life systems of the earth, we have no science that deals with the integral functioning of the earth itself. The term "Total Earth Science" suggested by Muller might be appropriate. However awkward this term might appear, it does have a certain precision in designating an area of understanding that has never been properly identified by our scientific establishment.

This lack of clarity in our fields of study seems to be one reason for the difficulty of humans finding their place within the dynamics of the earth system and why dealing with this subject has been so problematic. Some beginnings have been made in studies of the macrophase dimensions of the earth in terms of its geosphere, hydrosphere, atmosphere, biosphere, nousphere, and their relations to each other. The studies of James Lovelock and of Lynn Margulis are most significant in this context. Yet their work is only a beginning. As with scientists, generally, students of the earth sciences are good in the analytical phase of what they are doing. However, they seem to be consistently less competent in their capacity to deal with the more comprehensive under-standing of their subject.

When we consider the integral functioning of the planet, we realize that the nonhuman species discover their survival context with only limited disturbance of the larger complex of life systems. They discover their niche rather quickly or else they perish. A certain stability seems to eventuate: a new equilibrium comes into being, a functional relation of things with each other.

In discovering its own niche, the human seems to be different from other species. The human must do with critical understand-ing what other species do instinctively. The difficulty is that hu-mans are genetically coded toward a further transgenetic cultural coding whereby the human establishes itself in its specific mode of being, a cultural mode of being handed on by postbirth education. Through this cultural mode of its being, the human establishes its niche with its own mode of interaction with other species. This is achieved both by instinct and by reasoned choice on a more extensive scale than is the case with other modes of being. The realm of freedom found in other modes of being, their cultural coding, is less highly developed than in the case of the human.

The difficulty here is primarily in the cultural order. We could have been genetically coded within a more determined mode of being that would not permit the destructive relations between the human and the nonhuman such as now exists. But that would inhibit the full responisibility of the human. It is precisely in the critical capacities of the human and in the associated freedom of choice where the difficulty lies. In this context cultural distortions become possible, distortions that occur in a unique manner with the human and that endanger both the human and the biological processes of the earth. Reason is less reliable than instinct.

The ecological crisis we are experiencing is a consequence of the fact that the geobiological niche of the human is not being fulfilled with the precision that characterizes the niche of the various insects, or the niche of the ants, or the niche of the wolf or the deer—and certainly not as surely as the niche of the trees or the flowers in their respective functions. In all these cases there is an identifiable context of activity that is fulfilled in a consistent manner, a certain role within the larger community of life.

In the nonhuman order any failure of adaptation might quickly lead to extinction of the disturbing species. The human, however, through its intellectual capacities, is sufficiently cunning that it can, for a period of time, subvert the forces that might normally lead to extinction. What the human cannot do is to avoid the degradation in its own mode of being that occurs with a certain inevitability when it fails in fulfilling its niche in the larger earth community. The difficulty has existed at least since the rise of the Neolithic, possibly in the later phase of the Paleolithic. Even from this early period, the stress from the human mode of being has been making itself felt in the nonhuman world, but as yet is within manageable dimensions.

The general biological law is that every species should have opposed species or conditions that limit each species so that no single species or group of species would overwhelm the others—something that would assuredly happen if even a bacterium were permitted to reproduce without limitation over an extended period of time. The difficulty with the human is that it must be self-limiting since by virtue of its intellectual insight and the extent of its freedom of choice, the human can disrupt the limiting factors within itself and in its relations with other modes of being.

This issue of limits applies to the very beginning of the universe, for if the original expansion forces of the universe had been unopposed by some other force, then the universe from the beginning would have exploded and would have dispersed into chaotic fragmentation. The basic cosmic force limiting the expansive force is gravitation. At the same time, the expansive force of the universe is limiting the containing force. The creative balance achieved in this earliest period and maintained throughout the vast period of time that the universe has been in existence is among the deepest mysteries of the universe. The mutual limitation achieved by these two forces is the prototype of all the creative tensions that enable the universe to be as ordered as it is within a world of such turbulence as we observe.

So too with every species. The dynamism of the creative forces bringing a species into being must submit to the limiting forces that enable a basic pattern of order to impose itself. Species come into existence not as individuals but as populations or communities within the larger complex of communities. The diversity is absolutely important and yet it remains a limited diversity, a diversity of interacting species caught up in a complex of life systems that are mutually limiting.

The situation is even more complex when we consider that no enduring equilibrium is ever achieved. The universe is always in a state of balanced imbalance, of creative turbulence. The number and kinds of species is constantly changing. Species are constantly coming into existence and passing out of existence, generally over such lengthy periods of time, however, that there seems to be a pervasive serenity throughout the entire order of things. The total causes for this we cannot identify. Nor is it likely that we will ever be able to identify these causes in any comprehensive manner.

Although we can acquire considerable knowledge about ourselves and about the ambient life communities, we cannot identify completely our own status as a species, which is becoming the single most critical issue in the realms of both human knowledge and human action. In the Western world we are in the process of revising our sense of history by including greater reference to the nonhuman context and consequences of human affairs. While we have in recent times understood history in terms of human and interhuman causes, we have never adequately dealt with the determinations of human activities that arise from the earth itself. Nor have we identified in any adequate manner the story of the universe, of earth, and of life as our own primordial history—or the story of the human as intelligible only within this process.

The Great Story of the universe, the story of earth, and the story of the human have seemed too distant from each other to be encompassed within a single presentation. Now, however, the urgency is such that we can no longer isolate these various components of the earth from each other. This urgency includes all these elements of the Great Story, but especially the story of the earth: its difference from the other planets; its primordial shaping; the emergence of life in the seas and on the land in all their varieties and their interdependence; the diversity in bioregional life communities from the arctic tundra to the tropical rainforests, from the high mountain communities to the coastal regions.

As we turn toward such comprehensive consideration of the earth, we find that one of the central disciplines, so much neglected in general education in recent times, is geography. Geography brings together in its own unique manner all the spheres of the earth: the watersphere, the landsphere, the airsphere, the life sphere, and the mindsphere. Geography's more immediate function, however, is to serve as the interface between geology and biology. When ecologists talk about bioregionalism they are dealing with what might be described as a regional life community to be understood primarily in terms of a specific geographical setting.

Indeed these terms, ecology and geography, were brought together as early as 1923 in the presidential address of H. H. Burrows to the Association of American Geographers, an address entitled "Geography as Human Ecology." Even earlier the Berlin geographer Ferdinand von Richthofen (1833–1905) in his teaching was communicating a study of regional structures of the earth and their interrelations.

This awareness of the interaction of the various regions of the earth and human development goes back to the fifth-century B.C. world of Herodotus and Thucydides, who dealt with the Greek world with significant reference to the geography of the region, and to Strabo (63 B.C.–A.D. 24), who described the influence of geographical conditions on Europe, Asia, and Egypt. Little was done to advance these beginnings in the later Roman period or in the medieval period. The cultural and political strife throughout these centuries was too intense. In the sixteenth century Jean Bodin (1530–1596) wrote his *Method for the Easy Understanding of History* in 1561 with a concern for geographical influences on history. This concern for geography as a pervasive influence in human cultural development had an extension in the work of Charles Louis Montesquieu (1689–1755), Johann G. von Herder (1744–1803), Alexander von Humboldt (1769–1859), and Henry Thomas Buckle (1821–1862), who carried this influence on into the nineteenth century.

Later Karl Ritter (1779–1859), from a more academic perspective, showed the influence of geographical features on human development. Then Friedrich Ratzel (1844–1904), with his extensive commitment to environmentalism, came to his views from the study of geography as a new scientific discipline. These are the sources from which geographical studies emerged in America at the end of the nineteenth century and which influenced a new generation of geographers such as Carl Sauer.

In all of these instances the study of the earth in a geographical context was carried out within an academic discipline, with little concern for the application of the study for the practical functioning of the human within the larger dynamics of the earth process. These were not scientists exploring for minerals for their economic values, nor were they trying to direct political decision making, nor were they trying to establish an integral earth science that would guide the larger dynamics of human cultures toward a more viable relation with the integral earth functioning: their purpose, much more limited and more detached, was academic understanding.

Political theories based on geographical considerations emerged in the 1920s through Sir Halford John Mackinder (1861–1947). He established geographical studies in England as an academic discipline at the London School of Economics while developing his theories of Eurasia as the geographical heartland of the earth and of history. In these times geopolitics was a theory to be used for political domination. Karl Haushofer's (1869–1946) teachings concerning the domination of the heartland as the primary step toward domination of the world, both as a mode of intellectual understanding and as a guide in political-military strategy, was used by the National Socialists in their program for territorial conquest.

So too in economic theory the study of the regional distribution of natural resources was developed as a basis of economic development. Allied with research projects aimed at discovery of the specific location of earth's resources, economic geography has advanced as a basic ancillary discipline to the more integral understanding of economics. The discovery of different ores and especially the discovery of petroleum have largely determined the course of both human and planetary history in the twentieth century. If economic geography serving the purposes of human exploitation of the planet were to be altered into ecological geography for the purpose of identifying the proper niche of the human within the larger purposes of the earth community, then a great advance might be made toward achieving a viable planetary system.

Earlier civilizations came into being in regions especially favored for developing agriculture on an enlarged scale for an increasing population. River valleys with abundant water where animals would come for bathing and drinking were just such places where human settlement could take place with a certain ease. Geography explains, to some extent, the role of the Nile, the Tigris-Euphrates, the Indus, and the Yellow River valleys in determining human affairs. This experience of the past needs to be incorporated into our programs for effective habitation of the planet in the future.

Interference with the biological cycles has come about through irrigation of land areas, selection of seeds, domestication of animals, and clearing of forests. Human increase continued throughout the centuries, such that by two thousand years ago there were

probably some three hundred million humans on the planet, occupying territory on every continent. Problems of pollution were
limited and local during earlier centuries, although the use of
land throughout the planet was already causing serious difficulty
as long as three thousand years ago.

The culmination of these trends in the twentieth century has
produced what might be considered the most radical confronta-
tion of the human with the other life systems of the planet. The
processes of natural selection have led to what might be indicated
as cultural selection. But back of it all there arises the question of
earth: how does the planet function and how does the human
fulfill its role in the planetary process?

I am suggesting here that the answers need to be sought to
some extent in geographical studies, with attention given to bioregional integration and especially how humans might become present to each bioregion of the earth in a mutually enhancing
manner. One of the reasons for lack of attention to geography
seems to be a suspicion of an implicit geographical determinism
in the cultural shaping of the human. This is unfortunate since
geography is one of the basic integrating disciplines for those who
would enter into ecological studies with their emphasis on the
single community that humans form with the earth and all its
component members.

General ecological studies can be too abstract or too theoretical
to constitute a recognized scientific discipline. Biological and geological studies can be too specialized. Environmental ethics is a
much needed study, yet it cannot proceed in any effective manner
without a larger understanding of the natural world. The more
humanistic realms of poetry and the natural history essay are
important to establish the emotional-aesthetic feeling for the wonders of the natural world and to awaken the psychic energies
needed for dismantling our present destructive technological-in-
dustrial-commercial structures and creating a more benign mode
of economic survival for the entire earth community. But these
humanistic insights are themselves mightily enhanced by a more
thorough understanding of the identifying features and intimate
modes of functioning of bioregions.

None of these studies can be done in isolation from the others.
Each is derived from and leads back to the others. The relationship of humans to the earth requires all these modes of inquiry,
all these modes of expression. Yet the total earth science indicated
here needs to be something more than a composite of these multiple ways of viewing the earth. For the present this idea of a total,

or integral, earth science seems implicit in what we presently designate as ecology. Another term coming into use is "Earth Literacy," a valid term denoting a basic concept needed throughout the educational program from the earliest years up through professional levels. Each of these terms has its own special value. My own expectation is that the study of ecological geography will have a significant role to play in the future as one of the most effective integrating disciplines leading to a total earth science.

Cosmogenesis

Brian Swimme
California Institute of Integral Studies

THE universe flared forth fifteen billion years ago in a trillion degree blaze of energy, constellated into a hundred billion galaxies, forged the elements deep in the cores of stars, refashioned its matter into living seas, spouted into advanced organic beings, and spilled over into a conscious self-awareness that now ponders and shapes the evolutionary dynamics of earth.

These are, as scientists like to say, the "facts." But what do we do with them? How does the fact of cosmogenesis—this great fact that is the culminating achievement of millions of humans laboring now for a hundred thousand years, developing technical languages, amassing empirical details, pursuing the bewitching promise that we would come to know the mysteries of the universe—how does the discovery of cosmogenesis affect human consciousness?

I think those philosophers are correct who predict that over the next few centuries this knowledge will work its way deep into the strata of human consciousness and will blossom into the fourth great mutation of Homo sapiens.

But I could be wrong. And many thoughtful people think I am. It's embarrassing for me now to admit that it took me many years to take their dismissals seriously. Awash in the study of cosmogenesis, energized by the astounding news of how the universe and life and human consciousness emerged, I was simply too fascinated with this story even to notice those who were not.

I do remember puzzling over Malcolm Muggeridge who announced one fine day that "no matter what scientists discover about the universe," nothing would change his fundamental worldview, which happened to be an evangelical form of Christianity. Apparently (though at the time I was not capable of believing this), an intelligent person could remain unmoved by the empirical discovery of the universe's emergence into existence.

Another encounter that forced me to reflect took place when

I was presenting some of the details of cosmogenesis at a conference in Santa Barbara. My single theme was that a new form of consciousness was emerging, one grounded in the story of cosmogenesis—the story of a universe beginning with the so-called Big Bang and unfolding through fifteen billion years to become this Great Community we find ourselves within. Soon after my presentation the psychologist James Hillman took the podium and announced: "I don't give a hill of beans about the Big Bang. In fact, I care more about beans than I do about the Big Bang!"

Nor could I conveniently chalk up such dismissals as resulting from the inadequate science education in the United States. My third and final example involves a preeminent physicist, Steven Weinberg, a Nobel laureate. Weinberg ends his book on the birth of the universe with this memorable line: "The more the universe seems comprehensible, the more it also seems pointless."[1] Now this was truly baffling. Here was someone who knew everything I did about the birth of the universe and yet he concluded that such knowledge was without value, or that its only value was in the assistance it could provide us in reaching the conclusion that the universe was meaningless.

Why such a disparity of responses to this great fact?

Natural scientists such as myself are prone to examine anything but their own consciousness. This comes from a peculiar attitude common to our profession that is convinced the facts are what matter. Not just that the facts are important. We have a deep faith that if we can just figure out what the facts of the matter are, the way to proceed will be clear. Perhaps it was such confidence that enabled me to remain convinced, for two decades, that the empirical facts of cosmogenesis—all by themselves—were enough to launch our species into a new understanding. I never took seriously the notion that this great transformation of consciousness might involve something within, something having to do with experience.

The question that intrigues me now is this: what is it that happens that leads a person to regard cosmogenesis as not just a scientific theory, or a string of empirical facts, but as a way of life: as a religious attitude that enables a fresh, and creative, and cosmological orientation within the world?

I don't have the answer to this vast question, but I do have a story that might shed some light on it.

This took place on Halloween. I was wandering with Denise and our two sons through the dark streets. Near the end of the night, worn-out, I decided to let the three of them go ahead and fleece the remaining houses on the cul-de-sac. To ease the pain in the small of my back, I crouched down close to the ground. There it happened. For just one moment, I left "California" and entered the "universe."

Because I do happen to live in California, a home of extraterrestrial abductions, endless spiritual channeling, and a hundred other highly imaginative and exotic experiments in consciousness, I need to state at once that I am speaking here of the most ordinary thing. I was just crouched down in the middle of the street. In one sense, nothing at all extraordinary took place. I was just crouched there.

It had been raining softly all evening. I was close to the ground and I noticed the thousand edges of black asphalt glittering with soft points of light from a street lamp overhead. For some reason all my roles seem to vanish, or drop out of sight. I was no longer "Dad," or "husband," or "college teacher," or "Californian." I was not primarily in a "city." I was not part of a "holiday" of the "United States." I was suddenly just there, just this being, breathing, in the mist, rained upon, thinking.

Such tremendous things had to happen before I could crouch there. Vast galactic storms. Explosions so violent they were beyond the reaches of the human imagination. Subtleties of cellular electricity in a quintillion prokaryotes. I could not recount them all consciously, but I could feel them in a sense; I rode upon them as if I were treading water at midnight in the middle of an ocean with a deep current flooding up from below me. I sat crouched, amazed at the story that we now know, and amazed at the story of greatness that we would never know, that would stay hidden forever in the chasms of the past, but a story that had even so unfurled over billions of years into this moment, this experience, this present.

I was stupefied to reflect that this being that I was could have found itself in so many different forms. This breathing thing could have crouched in the bush of the great plains while its bonded mate gave birth to a new face off in the darkness. Could have huddled starving outside a cave that had been snowbound for thousands of years. This here-now could have been inside the night of a peaceful Paleozoic swamp, or screaming with reptilian terror in an ancient tropical forest. And this crouched and breathing being was now surrounded by those same opaque mysteries that, for unfathomable reasons, had taken on the form of asphalt.

Something moved. Crawling into my moment was a narrow black insect. She too was suddenly just there. I didn't know her name, but she was drenched. She was moving very slowly. To me she seemed hesitant, or confused. She changed directions uselessly.

Crouched down, I met her on a primordial level. I found myself thinking that she would never comprehend "asphalt," or "California," or "humanity." I sat staring, thinking how she was unable to read up on the situation. Unable to learn how to deal with roads, unable to learn how humanity had emerged and had flung up a thousand changes impossibly beyond her capacity for understanding. Here before me was a mentality of astonishing capabilities, one brought forth through a four hundred million year creative adventure, but a mentality unable to grasp the meaning of finding herself in this sea of "asphalt."

I had learned years before that through some supremely mysterious process the molten lava of earth had, over four billion years, transformed itself into organic life; but in that moment, crouched in the rain, I experienced this understanding directly. It had been decades since I had first learned that the same genetic language informed humans and insects, but now I could feel how I and this insect had been woven out of those same intelligent patterns. It had been so long since I first learned that all the elements of our solar system had been created in a star and then scattered by a supernova blast five billion years ago, but here in the night I experienced directly how I shared a common flesh with my confused and struggling kin.

Of course, I realize now how vulnerable to ridicule all this is. What fun a New York Times–type could have with this scientist going soft over an insect's plight! But in the moment such thoughts were the furthest thing from my mind. In the moment, I felt that I had suddenly understood, once again, the inner meaning of all our scientific studies.

The discovery of cosmogenesis is the discovery of a way of entering a more profound relationship with our cousins of this great community. On occasion, in an entirely haphazard manner, I have found my way into a brief taste of this experience. Haphazard because these moments are not planned; haphazard because in each such experience it feels as if it were for the first time, so effectively do I erase the memories of these experiences.

I end these reflections with two questions. First, is it the case that cosmogenesis remains just a scientific theory for those who have never experienced directly its truth? And second, is it possible that when humans begin consciously to create educational forms trained on evoking an embodiment of cosmogenesis, they will be activating the next era in the evolution of earth?

Note

1. Steven Weinberg, *The First Three Minutes: A Modern View of the Origin of the Universe* (New York: Basic Books, 1977), 154.

Notes on Contributors

THOMAS BERRY is a teacher, a historian of religion, and director of the Riverdale Center for Religious Research in New York City. As a "geologian," he is one of the leading spokespersons for a cosmological evolutionary worldview that evokes a sense of the spirituality of the earth. His books include *The Religions of India, Buddhism, The Dream of the Earth,* and, with Brian Swimme, *The Universe Story.*

BRIAN BROWN is an associate professor of religious studies at Iona College. He is the author of *The Buddha Nature: A Study of the Tathagatagarbha and Alayavijnana* (1991). He also writes and lectures on the religious and legal implications of environmental issues.

NOEL J. BROWN is the director of the Regional Office for North America of the United Nations Environment Programme. He has been a visiting professor at the University of the West Indies, City University of New York, and St. Mary's College of Maryland, and a distinguished lecturer at the University of Victoria, British Columbia. He was active in the preparatory process leading to the 1992 Earth Summit where he was a representative of the UNEP. Since then he has been exploring innovative ways to facilitate rapid implementation of Agenda 21, the Summit's Blueprint for a Sustainable Future.

J. BAIRD CALLICOTT is a professor of philosophy and natural resources at the University of Wisconsin, Stevens Point. He is the author of *The World's Great Ecological Insights: A Global Survey of Traditional Environmental Ethics from the Mediterranean Basin to the Australian Outback* and *In Defense of the Land Ethic: Essays in Environmental Philosophy.* He edited the *Companion to A Sand County Almanac,* and co-edited, with Susan L. Flader, *The River of the Mother of God and Other Essays by Aldo Leopold.*

CHRISTOPHER KEY CHAPPLE is an associate professor of theology at Loyola Marymount University in Los Angeles. He has published several articles on South Asian religions and is the author of

Karma and Creativity and *Nonviolence to Animals, Earth, and Self in Asian Traditions.*

DAVID RAY GRIFFIN is a professor of philosophy of religion and theology at the School of Theology at Claremont and Claremont Graduate School. He is the author of *God, Power, and Evil* and *Evil Revisited;* co-author of *Founders of Constructive Postmodern Philosophy;* editor of *The Reenchantment of Science* and *Sacred Interconnections;* and co-editor, with Richard Falk, of *Postmodern Politics for a Planet in Crisis.*

JOHN A. GRIM is an associate professor of religion at Bucknell University, where he teaches courses on Native American religions and indigenous religions. He has done field work with the Crow Indians in Montana and with the Kettle Falls people on the Columbia River in Washington. He is the author of *The Shaman* (1983). At Bucknell he was the principal coordinator of a yearlong program of speakers, workshops, and exhibitions investigating the significance of the Columbian Quincentennial during 1992.

ERIC KATZ is an assistant professor of philosophy and director of the Science, Technology, and Society Program at the New Jersey Institute of Technology and vice-president of the International Society for Environmental Ethics. He has published two annotated bibliographies of the field of environmental ethics and many essays on environmental philosophy. In 1990 he was a guest lecturer in the Lehrhaus program, "God, Man, and Nature in Judaism," at the Jewish Theological Seminary. He is currently writing a book on deep ecology.

JAY McDANIEL teaches environmental theology and world religions at Hendrix College, where he is an associate professor of religion and director of the Steel Center for the Study of Religion and Philosophy. He is the author of *Of God and Pelicans: A Theology of Reverence for Life* and *Earth, Sky, Gods, and Mortals: Developing an Ecological Theology* and co-editor of *Liberating Life: Contemporary Approaches to Ecological Theology, Good News for Animals?: Christian Approaches to Animal Well-Being* and *After Patriarchy: Feminist Transformations of the World Religions.*

RALPH METZNER is a psychotherapist, a professor at the California Institute of Integral Studies in San Francisco, and president of the Green Earth Foundation in El Verano, California.

tion, and *Learning and Politics: Essays on the Confucian Intellectual.* A member of the Committee on the Study of Religion at Harvard and a fellow of the American Academy of Arts and Sciences, he is currently interpreting Confucian ethics as a spiritual resource for the emerging global community.

MARY EVELYN TUCKER is an associate professor of religion at Bucknell University, where she teaches courses in world religions, Asian religions, and religion and ecology. She is the author of *Moral and Spiritual Cultivation in Japanese Neo-Confucianism* (1989) and various articles on the relation of religion to ecology. She specializes in the study of Confucianism in Japan. She is a committee member of the United Nations Environmental Programme for the Environmental Sabbath, an associate of the Global Forum of Spiritual and Parliamentary Leaders, and a vice president of the American Teilhard Association.

ROBERT A. WHITE holds a bachelor's degree in agriculture from the University of Saskatchewan and a master's degree in environmental studies from York University. He is currently pursuing doctoral studies in environmental education at the University of Toronto. He is also a consultant to the Office of Environment of the Baha'i International Community in New York.

LARRY L. RASMUSSEN is Reinhold Niebuhr Professor of Social Ethics at Union Theological Seminary in New York and co-moderator for Unit III of the World Council of Churches, "Justice, Peace, Creation."

GEORGE SESSIONS teaches philosophy at Sierra College and has written extensively in ecophilosophy and deep ecology. He is the co-author of *Deep Ecology* (1985), co-editor of *Environmental Philosophy* (1993), and editor of *Deep Ecology for the Twentieth Century* (forthcoming).

CHARLENE SPRETNAK's work has contributed to the framing of the women's spirituality, ecofeminist, and Green politics movements. She is the author of *States of Grace: The Recovery of Meaning in the Postmodern Age, The Spiritual Dimension of Green Politics,* and *Lost Goddesses of Early Greece;* co-author of *Green Politics;* and editor of the anthology, *The Politics of Women's Spirituality.*

BRIAN SWIMME directs the Center for the Story of the Universe, a research affiliate of the California Institute of Integral Studies in San Francisco. He is the author of *The Universe is a Green Dragon, The Universe Story* (with Thomas Berry), and the video series *Canticle to the Cosmos.*

ROGER E. TIMM is pastor of St. James Lutheran Church in Naperville, Illinois. From 1989 to 1992 he was college pastor and associate professor of religion at Carthage College, and from 1980 to 1988 he was assistant professor of religion at Muhlenberg College, where he first became interested in the environmental implications of Islamic creation theology.

MICHAEL TOBIAS is the author of fifteen books, and the writer, director, and producer of over sixty films. His best-known works include the novel and ten-hour television miniseries, *Voice of the Planet,* and the PBS film about Jainism, *Ahimsa: Non-Violence,* with its accompanying volume, *Life Force—The World of Jainism.* His other works include a play, *Harry and Arthur,* and the novels *Believe* and *Fatal Exposure.* His latest novel is *Mahavira.*

TU WEI-MING is a professor of Chinese history and philosophy at Harvard University. He is the author of *Neo-Confucian Thought: Wang Yang-ming's Youth, Centrality and Commonality, Humanity and Self-Cultivation, Confucian Thought: Selfhood as Creative Transforma-*